The Future of Association Leadership Starts Here

AI FOR ASSOCIATION LEADERS

The Essential Playbook
for CEOs and Boards
Ready to Scale Smarter,
Serve Better, and Lead the Future

ANNIE GIBBINS

Annie Gibbins / Women's Biz Global

Sydney, NSW Australia

Annie Gibbins Speaker - www.smartassociations.ai
Annie Gibbins Mentoring - www.leadwithannie.com
Women's Biz Global – www.womensbizglobal.com

AI for Association Leaders / **Annie Gibbins**. -- 1st ed.

ISBN 978-1-922969-24-8

DEDICATION

For every association leader who's been told to "wait their turn," and chose instead to lead the evolution.
This is for you.

ACKNOWLEDGEMENTS

Writing a book is never a solo journey. It takes courage to begin, stamina to continue, and the support of extraordinary people who walk alongside you, cheer you on, challenge your thinking, and help you cross the finish line.

This book was born from decades of experience in leadership, but it was brought to life through the energy, encouragement, and wisdom of many.

First, to the board directors, CEOs, practitioners, and changemakers I've had the honour of working with, you are the inspiration for this book. Every question you asked, every challenge you shared, every moment of vulnerability or boldness… it's all here. Thank you for your trust, your openness, and your willingness to lead with heart.

To my incredible teams at the Australian Traditional Medicine Society, Glaucoma Australia, the Australasian Society for Ultrasound in Medicine, and the Australasian College of Optical Dispensing, thank you for being part of the lived leadership that made this book possible. The work we did together was not always easy, but it was always meaningful.

To the Women's Biz Global community, you've shown me what's possible when passion and purpose collide. Your stories of resilience, reinvention, and ambition inspired much of the thinking behind this book's approach to smart, human-first innovation.

To my mentors and fellow thought leaders, thank you for modelling what it looks like to lead with integrity and clarity. I've learnt from the best, and I'm committed to passing that wisdom forward.

To my clients and collaborators in the association sector, thank you for being courageous enough to try something new, to test new ideas, and to show your industries what's possible when we let go of "the way it's always been."

To my behind-the-scenes creative and editorial partners, your ability to translate complex ideas into clear, compelling content is a gift. Thank you for your commitment to excellence, and for pushing this book to reach its full potential.

To my global community of leaders, innovators, and association professionals, thank you for showing up with curiosity. You are the reason this work exists. Keep questioning. Keep testing. Keep growing.

And finally, to my family, James, our amazing kids, and our beautiful grandkids, thank you for your constant encouragement, your unconditional belief in me, and your willingness to let me chase big ideas with relentless energy. You are my why. Always.

This book isn't just a guide.

It's a thank you.

To the people who believe in smarter systems, stronger leadership, and a future that works, for everyone.

Together, we are not just evolving.

We are leading the evolution.

With deep gratitude,
Annie Gibbins

FOREWARD

Artificial intelligence (AI) is the latest technical development in my professional career. Photocopiers, fax machines and computers arrived, then desktop computers which lay people could operate, voice dictation, and of course mobile phones. Already I have been amazed by what AI can do on a simple level at this early stage of its development.

For most of my working life, I have been involved in the association sector, and therefore I am impressed by Annie Gibbins writing a tome on the potential of AI for associations. **AI for Association Leaders** has been written to empower association CEOs and Boards to embrace AI in practical and strategic ways.

All associations are organisations with a purpose beyond profit for the owners – because we have members who cannot receive profits. However, the causes and the way entities operate vary greatly – and this is where artificial intelligence is a key new resource for association leaders.

Whilst **AI for Association Leaders** address the theory and ethics of AI, it also gives practical tips and techniques to implement ChatGPT and other tools. There is much content in his book which goes beyond AI matters to look at leadership and culture.

Overall, this book sends a positive message of AI as a tool for relevance, not just efficiency. Association Chairs, Presidents and heads of staff may gain greater confidence that they can utilise AI to lead their associations to advance the mission.

John Peacock AM

CEO, Associations Forum

AUTHOR'S NOTE

If you're holding this book, you're likely a leader who cares — deeply.

I wrote this book for you.

For the CEO navigating impossible expectations with limited resources.

For the board member wrestling with what "relevance" looks like now.

For the association team juggling spreadsheets, strategy, and the stories that matter.

I've been there.

For more than 15 years, I've led associations through change, growth, resistance, and reinvention — often at the same time. From Women's Biz Global to the Australian Traditional Medicine Society and beyond, I've seen firsthand the pressure, passion, and possibilities of association life.

And I know this: you want to do meaningful work that lasts. But the pace of change is fast. And the rise of AI? It can feel overwhelming, even destabilising.

This book is not about turning you into a tech expert. It's about helping you lead wisely, humanely, and with purpose — in a world where AI is no longer optional.

AI doesn't replace the heart of what we do.
It releases it.

So, let's move beyond the fear, the jargon, and the "one day" plans. Let's walk together — toward clarity, confidence, and tools that can actually help you build what matters.

I see you.
I get it.
Let's unlock what's possible — together.

— Annie Gibbins

CONTENTS

PART 1

Relevance in a Changing World

"AI isn't here to replace us; it's here to release us. It frees up our time so we can focus on what truly matters, connecting with and serving our members."

– *Annie Gibbins*

CHAPTER ONE

THE WAKEUP CALL

Let Me Begin With This: I See You.

I see you showing up every day with a mountain of responsibility and a calendar that never seems to shrink. I see you juggling member needs, Board expectations, budget constraints, and a growing sense that the ground is shifting beneath your feet.

You're not just running an organisation — you're stewarding a profession. You're holding up a cause, a mission, a vision that matters. You're working long hours to keep your community connected, your team focused, and your association relevant.

And I know — because I've been there too.

I've spent over 20 years leading purpose-driven associations. I've sat in the CEO seat of large, national bodies in health, education, and traditional medicine. I've led teams through restructures, rebrands, tech meltdowns, member revolts, and sudden market shifts. I've fought to deliver value with lean budgets, outdated systems, and Boards split between progress and preservation.

So I don't come to this topic from a distance. I come to it with sleeves rolled up, heart open, and deep respect for the real work you do.

Because here's what I know for sure: Association leadership is not for the faint of heart. It's layered. It's complex. It's lonely sometimes. And it's also one of the most important leadership roles you can step into — because associations don't just serve members. They shape industries. They carry legacies. They speak for people who often go unheard.

But right now, a reckoning is happening — and no association is immune from it.

Relevance isn't guaranteed — it's earned.

Let's start with the uncomfortable truth: Just because your organisation has been successful in the past doesn't mean it's relevant today.

In fact, your past success might be the very thing holding you back.

I've seen it time and time again — strong mission, committed team, loyal members… and yet the growth has stalled. The member engagement feels flat. The energy is low. You keep pushing, but nothing feels like it's landing. And when you raise the idea of change, especially something like AI, people flinch. Or freeze.

The truth? Relevance is not a legacy you inherit. It's a decision you recommit to every single day.

It's about staying obsessively connected to what your members need now, not just what they needed five years ago. It's about having the courage to evolve before you're forced to. It's about recognising that the biggest risk isn't moving too fast — it's staying still in a world that won't wait.

That's where this book comes in.

This isn't a tech manual. It's not about turning you into a coder or a chatbot expert. It's a leadership guide, written by someone who's led from the frontlines, to help you understand what AI actually is, what it can do for your association, and how to adopt it without losing the heart of who you are.

Because AI isn't a replacement for people.
It's a release.
A release from duplication, delay, and decision paralysis.
A release from the admin that burns out your staff.
A release from the bottlenecks that keep you stuck in survival mode.

And more importantly?
It's a return.
A return to purpose.
To connection.
To high-impact leadership that doesn't require burning out to prove your worth.

The Legacy Crisis – When Good Associations Fall Behind

Here's a truth that stings a little:
It's not just struggling associations that fall behind.

It's often the successful ones.
The respected ones.
The ones with a proud history, loyal members, strong reputations, and decades of tradition.

I've seen this over and over again — great intentions, great heritage, and… stagnation.
Why?

Because legacy, when left unexamined, can quietly become a limitation.

When Success Starts to Work Against You

Legacy is powerful. It's something to be proud of. You've built something that matters.

But legacy can also lull you into a false sense of security. It can whisper,

"We've always done it this way, and it's worked until now."

But *until now* is the key phrase.

We're in a new era.
One where expectations shift overnight.
Where members want instant access, personalisation, intuitive platforms, and content that cuts through the noise.

Legacy systems, rigid processes, and tradition-for-tradition's-sake won't cut it anymore.

A Real-World Example

A few years ago, I was invited to work with a well-established national association in the health space. They had history. Credibility. A long-serving Board. They also had:

- A 12-year-old website that wasn't mobile responsive.
- A CRM system nobody fully understood.
- An approval process that required three people and five days to send a member email
- A proud culture that was quietly eroding because new members didn't feel seen or heard.

On the surface, they were "fine."
But when we looked deeper, retention was down.

Member satisfaction had plateaued.
And new graduates were choosing smaller, more agile networks instead.

Why?

Because legacy had become their comfort zone.
And comfort zones are the most dangerous places to lead from.

What Got You Here Won't Get You There

You can have the best mission in the world.
You can believe deeply in your values.
You can have a team of incredibly hard-working people.

But if your tools, systems, and mindset are still operating like it's 2010, you're not in maintenance mode — you're in slow-motion decline.

And here's the worst part:
You might not even realise it until it's too late.
Because slow decline doesn't come with alarms.
It comes with little drops in engagement. Slight dips in revenue.
Staff that are "just tired."
And members that quietly drift away.

Don't Replace Your Legacy — Reinterpret It

Here's what I tell every association leader stuck in this trap:

- You don't have to burn it all down.
- You don't have to become a start-up.
- But you *do* need to be willing to evolve.

You can honour your past *and* build for the future.
But not if you're clinging to processes, platforms, or thinking patterns that no longer serve your purpose.

AI isn't about erasing what makes you great.
It's about helping you deliver your greatness in a way that actually meets the needs of this moment.

Your legacy isn't what's in your archives.
It's what you choose to do next.

The Culture of Busy – Where Good People Get Stuck

Ask anyone working in an association how they're doing, and you'll likely get the same answer:

"Busy."

Busy chasing invoices.
Busy replying to member emails.
Busy updating spreadsheets, logging CPD points, planning events, taking minutes, rewriting reports...
It never ends.

And because we're all so busy, we assume we must be productive.
But here's the uncomfortable truth:

Busy doesn't mean effective. And it definitely doesn't mean strategic.

In fact, the busier your team is, the *less capacity* they have to see what's not working, fix root problems, or create anything new.

Busyness Is a Badge — and a Barrier

Somewhere along the line, we started treating busyness like a badge of honour.

Long days. Back-to-back meetings. Overflowing inboxes.
We wear our exhaustion like proof of our value.

But what if all that motion… isn't actually moving us forward?

In many associations, busyness becomes the default culture. Teams fill their time with workarounds, duplicated effort, outdated tools, and systems held together by duct tape and goodwill.
They *feel* productive — but they're just spinning wheels.

You know the signs:

- Manual processes that could be automated
- Three people doing what one well-configured system could do.
- Time lost chasing approvals or reconciling inconsistent data.
- Good ideas that never get tested because there's "no time."

And here's the kicker: When you suggest innovation? The response is almost always…

"We're too busy for that right now."

I've Been There Too

In one association I led, we were flat out.
Every week was a sprint.
But when I stepped back and audited how we spent our time, the data stunned me:

42% of our team's hours were going toward admin-heavy, low-value tasks.
Tasks that could have been systemised, streamlined, or solved with basic automation — or even AI tools we already had access to.

We weren't lazy. We weren't inefficient.
We were just stuck in the loop of what we'd always done — too busy reacting to build anything better.

The moment we saw it clearly; we got to work.
Not on adding more.
On removing the friction.

We automated what we could.
Delegated what we didn't need to own.
Freed up time for strategic thinking.
And started looking at AI not as a threat, but as a lifeline.

AI Doesn't Just Save Time. It Gives Time Back to What Matters.

Let's be clear — I'm not romanticising technology.
I'm human-first, always.

But here's what I've seen firsthand:
When you implement AI in the right places, in the right way, your people get their energy back.
They stop drowning in tasks.
They start focusing on what only *they* can do — the creative work, the connection, the real service to members.

AI can:

- Draft standardised responses
- Prepare meeting summaries.
- Automate renewals, reminders, registrations.
- Personalise member emails at scale.
- Extract insights from surveys and databases.
- Reduce admin by **hours** per person, per week.

That's not just efficiency.
That's renewal.

When your team is no longer burnt out from doing the basics, they start to re-engage.
They bring new ideas.
They reconnect to purpose.
They lift the entire energy of the organisation.

The Real Cost of Busy

If you're too busy to innovate, too stretched to evaluate, or too overwhelmed to lead…
That's not sustainable.

You're not broken — but your system might be.

Busyness should never be a barrier to progress.
It should be a red flag that something needs to change.

And the good news?
It can.

But only if you stop accepting busy as the norm — and start building something smarter.

Where Friction Hides — And Why It's Blocking Progress

Let's talk about friction.

Not the loud kind.
Not the kind that creates scandal or sparks Boardroom blowups.

I mean the quiet, everyday kind.
The hidden drag that slows everything down without ever making a scene.

Friction in your association doesn't usually show up in strategic plans.
It doesn't raise red flags in your dashboards.

But you feel it. Your team feels it. And your members experience the ripple effects, even if they never say it out loud.

What Does Friction Look Like?

Friction is that event registration form that takes six clicks and two email confirmations.
It's the spreadsheet that only one person knows how to update.
It's the database that doesn't talk to your email platform.
It's the committee that needs to run everything through a subcommittee.
It's the annual report that gets rewritten five times because no one's quite sure what "final" means.

Here's what I've seen time and again:

Your team isn't inefficient.
They're navigating invisible resistance every single day.

And that resistance adds up — in hours, in morale, in momentum lost.

The Hidden Cost of Staying Stuck

Let me give you a real-world example.

I worked with an association where every single communication, from welcome emails to CPD reminders, was written manually by one team member. She was amazing. She cared deeply. But when she went on leave, everything ground to a halt.

There were no templates. No automations. No processes. Just heroic effort… and burnout waiting to happen.

That's friction.

And what's worse?
It often hides in places leaders aren't looking.

Because when you're focused on membership growth or policy advocacy or the next big event, you might not notice that your internal gears are rusting. Slowly. Quietly. But enough to keep you from moving forward with any real speed.

LEADERSHIP REFLECTION

Ask yourself:

1. Are you still approving social media posts through three people?
2. *Do you have legacy systems no one fully understands?*
3. *Are staff manually updating data across multiple platforms?*
4. *Is onboarding a new member (or employee) a confusing maze of emails and checklists?*
5. *Do great ideas get lost because there's "no clear process" for making them real?*

If the answer is yes to two or more — you've got friction.
And the good news? Friction is fixable.

But not if you ignore it.

AI Can't Fix Chaos — It Amplifies What's Already There

This is why I'm not someone who says, "Just use AI and everything will be solved."

Because if your systems are messy, your data is siloed, your workflows are vague, or your culture is fearful?
AI will just speed up the dysfunction.

It'll automate the wrong message.
It'll personalise junk.
It'll make assumptions based on flawed data.
It'll spit out answers that nobody trusts because nobody defined the question.

So before you introduce AI, you have to get real about what you're building on.

That's why this section is so important.

Because when you clear the friction first, AI becomes a multiplier — not a magnifier of your problems.

Tool-Smart Starts with System-Smart

You don't need more tools.
You need smarter systems.

And AI is one of those tools that shines when it's paired with simplicity.
Clear goals.
Clean processes.
Honest workflows.
Defined roles.

That's what allows AI to reduce pressure, not add to it.

So before you add, pause and ask:

- What's slowing us down?
- What's duplicating effort?
- What's taking energy that shouldn't?

The answers will tell you where your biggest breakthroughs lie.

From Fear to Forward Motion — Why Leaders Hesitate (And What to Do Instead)

Let's talk about the real reason many associations haven't embraced AI yet.

It's not budget.
It's not capacity.
It's not member appetite.
It's not even the technology.

It's fear.

And I say that with deep empathy — not judgement. Because I've felt it too.

When you've been leading in high-stakes, high-responsibility roles for years — the idea of stepping into a completely new way of working can feel risky. Vulnerable. Exposing.

You've worked hard to become competent. Reliable. Respected.

The thought of experimenting with something you don't fully understand?
It's uncomfortable.
What if I make the wrong call?
What if it fails?
What if I look like I don't know what I'm doing?

So you wait.
You research a little.
You send a team member to a webinar.
You bookmark an article.
You say, "Let's revisit that next quarter."

And all the while, the world moves forward.

The Language of Fear (That Doesn't Sound Like Fear)

In association leadership, fear rarely shows up as fear.

It shows up as logic:

- "We're not ready."
- "Our members aren't asking for it."
- "We don't want to overwhelm the team."
- "We'll review this as part of our next strategic planning cycle."

Sound familiar?

I've heard these statements from smart, capable leaders who care deeply about their organisations.
But beneath the surface? Often, there's a quiet sense of paralysis.
Because they don't want to break what's working — even if what's working… is barely holding on.

The Risk of Waiting

Here's what I know from experience:
The longer you wait, the harder it gets.

Not because the tools become more complex — but because the gap between you and your members, your competitors, your potential — it widens.

You don't have to do everything at once.
But you do have to start.

And here's the reframe that helped me most:

It's not about replacing people. It's about releasing potential.

When I stopped seeing AI as a threat to our values and started seeing it as a vehicle for our vision — everything shifted.

It wasn't about jumping on a trend.
It was about finally having the tools to scale our impact without burning out our people.

Start Small, Move Smart

If you're feeling overwhelmed, here's my advice:

Start tiny.

- Choose one process to automate.
- Use AI to summarise meeting notes.
- Test a chatbot to answer common member questions.
- Use it to draft a comms plan or brainstorm ideas for a campaign.

The key is movement — not mastery.

Every time you use AI in a way that saves time, sparks insight, or lifts a burden from your team, you build confidence.

You build proof.

You stop guessing and start growing.

Your Members Deserve a Leader Who's Awake

You don't need to be a tech guru.
You don't need to have all the answers.

But you *do* need to be awake.

To the pace of change.
To the expectations of your members.
To the tools now available that can help you serve with more heart, more clarity, and more ease.

You can't afford to sleep through this shift.

Because the cost of staying still isn't just efficiency.
It's engagement.
It's credibility.
It's relevance.

And you — your association, your mission, your legacy — deserve better than that.

LEADERSHIP REFLECTION

What's Really Going On Here? Take five minutes to sit with these questions before turning the page:

- *Where are we clinging to "how it's always been done"?*
- *What do I* know *needs to change — but I've avoided acting on?*
- *What part of our member experience would make me walk away if I were a member?*
- *If I had one day without meetings, what strategic change would I start?*

This is where momentum begins — with clarity, not capacity.

The AI Readiness Scorecard — How to Measure Your Momentum

By now, you've probably realised something important:

You don't need more urgency.
You need more clarity.

Clarity about where you're starting from.
Clarity about what's working.
Clarity about what's holding you back.

That's where the **AI Readiness Scorecard** comes in.

This isn't about giving your association a grade.
It's about giving you a mirror — so you can see, without judgement, what's really going on.

You can't optimise what you can't see.
And you can't transform what you're afraid to evaluate.

This scorecard is designed to help you pause, reflect, and understand how prepared your association really is to use AI — not as a gimmick, but as a growth tool.

What Are We Measuring?

We're not just looking at tech infrastructure.
We're looking at leadership, systems, culture, and readiness to adapt.

Because here's what I've learned:

It's not the lack of tools that stops most associations from evolving. It's the presence of friction, fear, and fatigue.

The Scorecard will show you where those things are hiding and how to move past them.

It will also highlight what you already have going for you.

I've worked with associations who felt stuck, only to realise they already had 70% of what they needed in place. They just needed the roadmap — and the courage — to connect the dots.

The 5 Dimensions of AI Readiness

You'll find the full scorecard in the appendix and also as a digital version you can share with your team. But here's the overview:

1. Leadership Alignment

- Are your CEO, executives, and Board aligned in understanding the role of AI?
- Do they share urgency and vision — or are they waiting for someone else to make the first move?

2. Operational Systems

- Are your systems connected, automated, and scalable or clunky, manual, and person-dependent?

3. Member Experience

- Do you know what your members need and expect — or are you assuming based on the past?
- Is their experience consistent, personalised, and friction-free?

4. Data & Insight

- Do you have access to clean, current, centralised data?
- Or are decisions being made by gut feel, opinion, or spreadsheet scavenger hunts?

5. Culture & Capability

- Is your team curious, open, and willing to learn — or burnt out, overwhelmed, and wary of change?
- Do you have time, training, and trust to experiment?

How to Use the Scorecard

You can use this tool in a few ways:

- As a **self-assessment**
- As a **leadership team discussion starter**

- As a **Boardroom prompt** to guide strategic planning

Each category includes prompts, ratings, and next-step guidance.

The point isn't to be perfect — it's to be honest.
Because when you know where you're starting, you can actually build momentum that lasts.

And you might be surprised — your association could be more ready than you think.

You Might Be Thinking…

"But our members are older — they don't want AI."
Actually, they want ease. They want answers. They want responsiveness. AI helps you deliver that — without making your systems feel impersonal.

"We can't afford to make a mistake."
The real risk isn't trying something new — it's letting your relevance slowly erode because you were too cautious to act.

"We've already got so much tech."
That's exactly why you need AI. To make it all talk to each other. To make it *work for you* instead of you working around it.

Progress, Not Perfection

Let me say this clearly:

- You do not need to have it all figured out.
- You don't need to overhaul everything this quarter.
- You don't need to fix what's not broken.

But you *do* need to start asking better questions.
You *do* need to open the door to smarter tools.

You *do* need to prepare your people, your systems, and your mindset for what's next.

That's what the rest of this book is here to help you do.

Let me give you a quick story of what's possible.

A regional industry association I worked with used to manually enter every member renewal by hand. Their team was exhausted, morale was low, and their new member onboarding process. Non-existent.

We introduced a simple AI-powered automation — a chatbot that handled member FAQs and triggered onboarding tasks. Within 30 days, their response time dropped from 72 hours to under 12. Their renewal process became instant. And most importantly?
The team had space again. They used it to redesign their value proposition.
Now they're growing. Not because they hired more people — but because they started thinking differently.

If Chapter One was your wake-up call, Chapter Two is your mirror.
It's where we stop talking about change in theory and start uncovering where associations are falling short in practice. We'll look closely at the disconnect between what leaders assume and what members actually experience.
Why? Because when we name reality, we unlock relevance. And from there, everything becomes possible.

Your Next Step

In Chapter Two, I'll walk you through **how to simplify your systems first** — so AI becomes a catalyst for clarity, not complexity.

We'll talk about where to begin.
What to focus on.
And how to set up your association to grow with confidence — not chaos.

But for now, here's what I want you to know:

You're not behind.
You're right on time.

You're not too late.
You're just getting started.

You don't need to lead with fear.
You get to lead with heart — and tools that actually support it.

"You don't need more time. You need more truth — and the tools to act on it."

CHAPTER TWO

FROM NOISE TO CLARITY

Forget Assumptions. Let's Talk About Reality

There's a dangerous phrase I hear far too often from well-meaning association leaders:

"Our members are happy with how we've always done it."

Let me tell you what I've learned:
By the time someone says that the members who weren't happy are already gone.

They didn't send an angry email.
They didn't fill out a feedback form.
They didn't lodge a formal complaint.

They simply left — quietly, invisibly, without making a scene.

Why?

Because they didn't feel heard.
They didn't feel known.
They didn't see their needs reflected in your decisions or your direction.

And the moment someone stops feeling seen is the moment they start disengaging — whether they're a brand-new member or a lifelong supporter.

That is the real crisis facing associations today.
Not just a membership crisis. Not just a revenue crisis.
A **relevance crisis.**

What Relevance *Really* Means

Relevance doesn't mean you're active on LinkedIn.
It doesn't mean you have a newsletter or a glossy conference brochure.
And it definitely doesn't mean you've been around for 50 years.

Relevance means this:
Your members see your organisation as critical to their growth, their confidence, and their future — right now.

Not last year. Not a decade ago.
Now.

And the only way to stay relevant is to **keep listening**.
Keep evolving. Keep asking better questions.

But here's the challenge: most leaders are too busy keeping the engine running to step back and check if they're even on the right road.

A True Story: The Silent Slide

Several years ago, I stepped into a senior leadership role with an association that had a glowing legacy.

It was loved by its founding members, respected in policy circles, and celebrated its heritage with pride.

But when I looked at the data, something wasn't adding up.

- Website traffic was declining.
- Event participation was dipping.
- Younger members were barely renewing.
- Social engagement was flatlining.

And when I asked the Board why, I heard things like:

- "Young people just don't engage like they used to."
- "We don't want to change what's working."
- "They'll come around once they see the value."

But here's the thing: they weren't coming around. They were logging off. They were drifting. And we were watching it happen in real-time.

The problem wasn't the members.
The problem was **our assumptions about the members.**

We thought we were still the centre of their professional universe.
But they had moved on — and we hadn't noticed.

You Can't Lead by Looking in the Rearview Mirror

Too many association strategies are built on:

- Last year's results
- Five-year-old survey data
- Anecdotes from the Board
- Fear of upsetting the status quo

That's not strategy. That's **survival mode** wearing a nice suit.

When you assume you know your members because you've known them in the past, you're setting your association up for irrelevance.

You don't get to define what's valuable. Your members do.

The question is: are you willing to listen — even when it challenges what you've always done?

Reflection for Leaders

Take a breath and answer these honestly:

1. When was the last time you asked your members what they actually need — and changed something because of it?
2. Are your events, content, and communications designed around your internal structure — or your member's lived experience?
3. What signals have you dismissed as "outliers" that might actually be early warnings?

This is where the transformation begins — not with a dashboard or AI tool, but with the willingness to **question your own certainty**.

Personal Belief from My Desk to Yours

After more than 20 years in leadership, I've learned this the hard way:

"The day you stop asking if you're still relevant… is the day you probably aren't."

Staying relevant takes courage.
It means letting go of outdated offerings — even if they've been beloved.
It means building for a future that hasn't arrived yet.
It means experimenting, listening, iterating — and yes, sometimes failing forward.

And you don't need a tech degree or a fancy budget to start.

You need one thing:
The humility to admit you might not know — and the bravery to find out.

The Shift from Content to Connection

Let's get real for a minute.

If you're running an association today, chances are you're producing more content than ever before.

You've got:

- Webinars happening monthly (sometimes weekly),
- eLearning modules,
- A growing library of CPD content,
- Monthly newsletters,
- Blogs,
- Podcasts,
- Thought leadership pieces,
- Resources and templates for every imaginable need.

And yet...
Engagement is dropping.
Click-throughs are flat.
Event attendance is inconsistent.
Members skim, delete, or ignore — even when the content is excellent.

It's not because your content is bad.
It's because it's not *connected.*

The Content Explosion Isn't the Problem. It's the Experience.

We are in the era of *content overload* — not content scarcity.

Your members can open their phones and access:

- World-class podcasts,
- Global experts on LinkedIn,
- Free downloadable tools,
- Niche communities answering their questions in seconds,
- AI platforms generating resources in real time.

The bar is not just high — it's global, immediate, and hyper-personalised.

So when your association sends out another 10-point newsletter or a "one-size-fits-all" email, it gets swallowed in a sea of noise.

This isn't a content crisis.
It's a **connection crisis**.

What the "Netflix Generation" Expects From You Now

And no — I'm not just talking about Gen Z or Millennials.
I'm talking about **everyone** who now lives in a digital-first, on-demand world.

They expect:

- Smart recommendations
- Content that's brief, digestible, and valuable.
- Curated journeys, not just libraries
- Platforms that remember them
- Notifications that actually help.

They don't want to search for what's useful.
They want it delivered before they even ask.

This is the world your members are already living in — and they're wondering why their association hasn't caught up.

A Case Study: Turning a Clunky CPD Archive into a Member Magnet

One of the associations I worked with had a brilliant collection of CPD content — over 200 hours' worth.

It was well-produced, peer-reviewed, and up-to-date.
But hardly anyone was watching it.

Why?

Because the content was organised like a dusty archive:

- Long dropdown menus
- Generic titles
- No filters
- No guidance

It felt like walking into a library with no librarian and no map.

Here's what we did:

- Used AI to categorise every resource by interest, career stage, and learning goal.
- Created pathways — e.g., "First 3 years in practice," "Private clinic growth," "Mastering compliance."
- Added "Next Best Step" prompts at the end of each module.
- Delivered personalised playlists to members via email based on their recent activity.

Within 3 months:

- Course completions doubled.
- Repeat visits tripled.
- Member feedback went from "overwhelming" to "it finally makes sense."

We didn't create new content.
We simply made it easier to access and feel relevant.

From Library to Journey

Here's the big mindset shift: **Your members don't want a content library. They want a learning journey.**

They want to feel like someone understands their needs, curates content accordingly and helps them take the next right step — not just dumps information in their inbox.

LEADERSHIP REFLECTION

Ask yourself:

1. *Is your content designed around internal departments or member goals?*
2. *Are you delivering information... or helping someone move forward?*
3. *Could AI help personalise your learning experience without adding work to your team?*

Because content without clarity = noise.
And clarity is what builds trust.

LEADERSHIP REFLECTION

Grab your notepad and ask:

1. *Are we measuring content success by output or impact?*
2. *Do we know what kind of content different member segments actually want?*

3. *How easy is it for a new member to find their ideal next step — today?*
4. *What could we stop doing right now that's cluttering the experience?*

My Challenge to You

You don't need to be Netflix.
But you can **learn from Netflix**.

You can design your learning pathways.
You can use simple AI tools to tag and recommend content.
You can stop flooding inboxes and start building journeys.

And you can be the association that doesn't just inform members —
You empower them.

In one mid-sized association I worked with, leadership was convinced that their members were "happy." Feedback surveys were positive. Events were moderately attended. No one was complaining.

But when we did a behavioural audit using AI, we discovered something quietly alarming: Over 40% of new members had disengaged within the first 90 days. No event registrations. No logins. No resource downloads.

They hadn't unsubscribed. They hadn't spoken up.

They'd simply... faded away.

When we asked why, the answer was sobering: "We didn't feel like this was built for us."

That changed everything. It made leadership stop asking "What do we offer?" and start asking, "Who do we serve—and what do they actually need?"

Belonging Is the New Loyalty

And You Can't Automate That

Let's pause the strategy talk for a second.

LEADERSHIP REFLECTION

Ask yourself:

1. *Why do people stay?*
2. *Why do they stay in a membership, a team, a movement?*
3. *Is it because they get a discount?*
4. *Because they receive a glossy magazine once a quarter?*

No. They stay because they **feel something**.

They feel understood.
They feel seen.
They feel like they belong.

That's the new loyalty.

It's not built on obligation.
It's built on identity.

When a Member Says "These Are My People"

I remember sitting at a networking breakfast a few years ago — just a low-key local event for new health practitioners.

It wasn't fancy.
No keynote speaker. No expensive venue. No tech.

But I watched something happen that I'll never forget.

A young member turned to someone she'd just met and said:

"I didn't realise how much I needed this. It feels like these are my people."

That moment.
That's your metric.
That's the gold.

Not the number of views.
Not the click-through rates.

Belonging is your most powerful retention tool.

Why Traditional Loyalty Has Collapsed

Associations used to rely on tradition:

- "You join because you're supposed to."
- "It looks good on your CV."
- "It's what professionals do."

But those norms are breaking down — and they're not coming back.

Why?

Because members today have options.
Free online communities.
Hyper-targeted LinkedIn groups.
Webinars from global leaders streamed on demand.

They can belong elsewhere — for free — and still get access to peers, support, and recognition.

So if you're relying on your legacy, your title, or your credential to hold people?
That's not loyalty.
That's **fragile familiarity**.

Belonging Isn't Something You Can Buy — or Automate

Here's the uncomfortable truth: You can't fake belonging.

You can't automate it.
You can't systemise it and expect it to feel real.
You can't create it once and assume it sticks.

Belonging comes from:

- Intentional spaces
- Shared stories
- Active inclusion
- Peer recognition
- Psychological safety

And none of that happens by accident.

The job of an association is no longer just to inform — it's to **connect**.

Case Study: Peer Circles That Changed Everything

One association I consulted with was struggling to engage their early-career members.
These were practitioners fresh out of training, nervous about setting up a business, unsure about regulations, and intimidated by established members.

The leadership team's idea?
Send more newsletters.

But we tried something else instead.

We created **peer circles** — intimate virtual spaces for first- and second-year members. No agenda. Just guided peer conversation and one senior mentor who listened more than they spoke.

We used AI to track participation and engagement trends, and nudged facilitators when someone had been quiet for a few sessions.

The result?

- Retention in that cohort jumped 27%.
- Attendance at events increased by 40%.
- Members started referring their peers to join.

Not because of benefits.
Because of belonging.

Belonging at Scale Is Possible — with the Right Support

Let me be clear — you *can* build belonging at scale.

You just need:

- Tech that removes friction
- Insight that helps you segment wisely.
- Personalisation that shows you care
- Humans who are freed from admin so they can show up.

AI can't create community — but it can create **capacity**.

It can:

- Flag disengaged members.
- Suggest timely interventions.
- Offer personalised messages.
- Free up your team to *do the human stuff.*

LEADERSHIP REFLECTION

Ask yourself:

1. *Who in your community feels invisible right now?*
2. *What groups have drifted because nothing speaks directly to them?*
3. *Are your spaces inclusive, accessible, and emotionally safe?*
4. *How can you enable more peer-to-peer connection — not just top-down content?*

What I Know for Sure

I've built membership models, redesigned engagement strategies, and tested dozens of campaigns.

Here's what I know:
People don't leave communities they feel deeply part of.

And the work of leadership in this new era is this:

- Stop trying to scale content.
- Start trying to scale *care.*
- Create space for people to see themselves — and each other — in what you build.

Belonging is what turns a membership into a movement.
That's your opportunity. That's your advantage.
But only if you choose to make it your priority.

From Insight to Action

What Your Data Knows That You Don't

Let me say this upfront:

Your association already has the answers.
The question is — are you paying attention?

Because data speaks.
Always.
Even when it's quiet.
Even when it's uncomfortable.

And more often than not, the insight you need isn't in the fancy dashboard or annual report — it's hiding in plain sight.

You just have to know where to look… and what to do with what you find.

Data Isn't a Report. It's a Compass.

One of the biggest myths I see in the association space is this:

"We don't have the data."

You do. You just might not be using it well.

Your data is in:

- The drop-off points in your webinar replays
- The bounce rates on your email campaigns
- The members who haven't logged in for 90 days
- The pages people visit before they cancel.

- The content that gets 3 clicks out of 3,000 recipients.

These aren't statistics.
They're *stories.*
Clues. Signals. Warnings.

And if you wait for the numbers to become alarming, it's already too late.

A Wake-Up Call: The Quiet Year-Two Dropout

One association I worked with had a big surprise waiting in their data.

Their new member campaigns were working.
Onboarding was strong.
Year-one engagement was healthy.

But they kept losing members in **year two**.

The team assumed the content wasn't valuable enough.
Or that they needed more perks.

But when we analysed behavioural data using a simple AI tool, the picture became clearer.

In year one, members were nurtured. They received regular touchpoints. Content was relevant and paced.

But once the first year ended, the journey stopped.
They were dumped into the general pool with generic updates and forgotten.

There wasn't a **problem**.
There was a **drop in care** — and members felt it.

We fixed it by:

- Creating a "second-year success track"
- Using AI to flag disengaged behaviour
- Triggering messages based on usage patterns.
- Re-onboarding at the 12-month mark

And just like that, second-year retention increased by **19%.**

That's the power of action-oriented insight.

Most Leaders Are Asking the Wrong Questions

They ask:

- "How many opened the email?"
- "What's our member growth?"
- "How many downloads did we get?"

But they should be asking:

- "What patterns are emerging in disengaged members?"
- "Where are people dropping off in their journey?"
- "Which segments are we ignoring — unintentionally?"
- "What small signals are telling a bigger story?"

AI can help you spot these.
Not with magic. But with **visibility**.

It can do what humans can't:

- Connect dots across hundreds of interactions.
- Detect early signs of churn.
- Recommend next best actions.
- Summarise engagement patterns in real time.

This isn't surveillance.
It's service.

Real-World Example: Predicting Churn Before It Happens

Another association I supported introduced a churn-prediction model.
Nothing fancy — just using an AI tool that scored member behaviour each month based on a few inputs:

- Logins
- Clicks
- Attendance
- Time on page
- Resource downloads

When a member dropped below a threshold, it triggered:

- A personalised re-engagement email
- A check-in from a community manager
- A new content path suggestion

Over 6 months, they reduced churn by **22%** — and their most at-risk members became some of their most engaged.

That's what happens when insight turns into *action.*

But Here's the Catch…

Most associations aren't doing this — not because they don't care, but because they're drowning in noise.

They're measuring everything except the stuff that actually tells a story.

Too many teams are stuck producing reports that nobody reads. Or chasing KPIs that don't change anything.

Insight is not about *tracking.*
It's about *transforming.*

LEADERSHIP REFLECTION

Ask yourself:

1. *Are we collecting data we actually use — or just storing it?*
2. *Do we know what disengagement looks like in our membership?*
3. *How quickly do we act when behaviour changes?*
4. *Could AI help us see what's happening — before it's too late?*

What I've Learned

You don't need more data.

You need **better questions**.
Smarter signals.
More meaningful measures of success.

And most of all, a willingness to act **before** the pain shows up in your bottom line.

Because when data becomes part of your culture — not just your reporting —
You don't react to problems.

You prevent them.

Designing a Member Experience That Feels Like It Was Built Just for Them

Let's imagine something together.

A member opens their inbox, sees a message from your association, and thinks:

"They know me."
"This is exactly what I needed."
"It feels like this was made just for me."

That's the kind of experience that doesn't just impress — it builds trust.
It creates loyalty.
And more than anything, it makes your association **essential** to that person's professional life.

The good news?
This is entirely achievable.
And you don't need a Silicon Valley budget to make it happen.

What you need is a shift — from delivering content *to everyone* to designing experiences *for someone.*

Why "Generic" is the Fastest Way to Irrelevance

Let's be honest: many associations are still treating all members the same.

Everyone gets the same:

- Newsletter
- CPD updates
- Event invites
- Survey requests

It's efficient.
But it's also invisible.

Because when something isn't tailored to me — I assume it's not *for* me.

Members don't want more content.
They want the right content.
At the right time.
In the right format.
Matched to their real needs, not just their membership category.

What Personalisation *Really* Means (and What It Doesn't)

Let's clear up a common myth.

Personalisation doesn't mean sending one-to-one custom emails.
It doesn't mean calling every member to check in.
It doesn't mean building 45 different versions of every resource.

It means:

- Segmenting based on behaviour, not assumptions.
- Offering relevant "next steps" based on recent activity.
- Allowing members to self-select what they want to receive.
- Using data and AI to recommend, not just deliver

It means *designing an experience* — not just distributing information.

Case Study: The First 30 Days Matter More Than You Think

I once audited a membership journey for a national association with high drop-off rates in year one.

When I signed up as a mystery member, here's what I got:

- A generic welcome email
- A password to a complex portal
- Zero contact for 3 months
- A renewal reminder at 11 months

No wonder new members weren't sticking.

We flipped the script.

We created a 30-day onboarding journey:

- A warm, human welcome video from the CEO
- A "choose your path" experience: new grad, business owner, researcher, etc.
- An AI-tagged email series based on their interests.
- A mid-month check-in: "How are you going so far?"
- A community invitation that matched their goals

Suddenly, members weren't just logging in — they were engaging.

Renewals in year one jumped by **31%**.

Not because we had more benefits.
Because we *designed a better experience.*

You Don't Need Perfection. You Need Precision.

Most associations try to do too much.

They want to:

- Serve everyone.
- Offer everything.
- Be all things to all members.

But in the process, they dilute their impact.

The better approach?

Start narrow.
Design a world-class experience for one key segment — and scale from there.

For example:

- New grads
- Solo practitioners
- Regional members
- Students transitioning to practice.
- Members running their first business.

Focus on what one group truly needs.
Deliver it well.
Use AI to test what works and adapt as you go.

The result?
You earn trust.
You create raving fans.
And you build a playbook for wider success.

A long-established national association representing thousands of complementary health professionals had all the traditional signs of success — a strong reputation, a clear mission, and steady membership numbers.

But something wasn't right.

Despite offering a growing suite of online CPD resources, webinars, and regular email updates, they noticed:

- Website bounce rates were rising.
- Event registrations were declining.

- Engagement from newer members had dropped off significantly.
- Year-two renewals were lagging behind expectations.

Internally, the team believed they had a content issue.
So they created more webinars, more emails, and added more resources to their member portal.

But none of it moved the needle.

Eventually, they used a lightweight AI tool to run a behavioural audit — not to gather more data, but to interpret what they already had.

What they discovered changed everything.

- First-year members were highly engaged… for the first 6 weeks.
- After that, unless they attended a live event or joined a community space, their activity dropped rapidly.
- Most members felt overwhelmed by content and unclear on what to engage with next.
- The CPD platform, though robust, was hard to navigate — especially for members accessing it via mobile.

Instead of launching another round of content, they made a smarter move:

- They simplified their onboarding with automated check-ins, goal selection, and personalised learning pathways.
- They used AI to track engagement and recommend "next best steps" to each member.
- They created peer pods — small, AI-matched discussion groups for members in similar career stages.
- They re-engaged at-risk members with targeted nudges based on behaviour, not guesswork.

Six months later:

- Email open rates jumped nearly 20%.
- CPD completions more than doubled.
- Year-two renewals increased by 22%.
- Word-of-mouth referrals increased from members who felt genuinely supported.

What shifted?

Not the content.
Not the budget.
Not the team size.

What changed was **how they listened — and how they responded**.

They didn't try to be everything to everyone.
They used the tools at their fingertips to **design experiences that felt relevant, personal, and timely**.

That's what modern association leadership looks like.

LEADERSHIP REFLECTION

Ask yourself:

1. *What do our first 30 days of membership actually feel like?*
2. *Are we solving real problems for real people — or delivering what's easiest to produce?*
3. *What would happen if we designed an experience just for one overlooked segment?*
4. *Where could AI help us personalise, recommend, or simplify — without more staff hours?*

What Experience Is *Really* About

It's not about flash.
It's not about tech for the sake of it.
It's about making people feel like they matter.

Because when someone says:

"They made this *for me*,"

That's when they stop being a member…
And start becoming an advocate.

Wrapping Up Chapter Two

If Chapter One was your wake-up call, Chapter Two is your compass.

You now know what members want:

- Relevance
- Recognition
- Connection
- Clarity
- Care

And yes, AI can help.
But the heart of this work is human.

Design experiences that see people.
Use tools that remove friction.
Build systems that *amplify* your humanity — not replace it.

In the next chapter, we're going to build your roadmap — the SMART AI Strategy™ — to help you lead with heart, but scale with smarts.

Are you ready?

Great! Because the future of your association won't be built by algorithms. It will be built by leaders who know how to use them — wisely, ethically, and with purpose.

"Change isn't comfortable, but neither is irrelevance. Choose your discomfort wisely." – Annie Gibbins

CHAPTER THREE

THE SMART AI STRATEGY™

Why Strategy Is Your New Superpower

Let's start with something most CEOs won't admit out loud:

- We want innovation — but we hate being vague.
- We crave results — but we dread reinventing the wheel.
- We love new ideas — but not without a roadmap we can explain to the Board by Friday.

Sound familiar?

That's exactly why so many leaders get stuck between ambition and action.

When it comes to AI, the noise is deafening.
Everyone's talking about it.
Everyone's experimenting with it.
And yet, the majority of associations I meet are frozen — unsure where to begin, how much to invest, or what success even looks like.

The solution isn't another app or tool.
It's not another webinar on ChatGPT.

It's certainly not hiring a consultant to write a 90-page transformation plan that no one reads.

What you need is a **framework that simplifies the chaos** — one that helps you move forward with clarity, confidence, and alignment.

That's what the **SMART AI Strategy™** is built for.

Why Most Digital Strategies Fail in Associations

Let's talk honestly for a moment.

When associations *do* try to get strategic about tech or AI, it often falls flat. Not because the intention is wrong — but because the **approach is flawed**.

Here's what I see time and again:

- The plan is too complex to execute with existing staff.
- It doesn't align with real member needs.
- The Board signs off — but no one truly champions it.
- The team gets excited for a month, then goes back to "business as usual."

You've seen this, right?

It's not just frustrating — it's exhausting.
And it reinforces the false belief that AI or innovation "just isn't for us."

But here's the truth:
It *is* for you.
You just need a way to approach it that's built around the realities of association leadership — not Silicon Valley hype.

Introducing the SMART AI Strategy™

The SMART AI Strategy™ is the framework I developed after years of guiding associations through digital transformation — without the overwhelm, without the jargon, and without losing what makes them human.

It's built on five pillars that reflect how associations **actually** work:

1. **Scalable** – You start small and grow with purpose.
2. **Measurable** – You define success clearly, from the start.
3. **Agile** – You adjust quickly, without waiting for perfect.
4. **Relevant** – You stay laser-focused on real member needs.
5. **Tech-Enabled** – You let the tools do the heavy lifting, not the people.

This isn't theory.
It's a strategic framework **you can implement immediately** — whether you have a 10-person team or a part-time admin and a volunteer Board.

SMART is designed to work within your current capacity and grow with your momentum.

Why Strategy Matters More Than Tools

Here's something you won't hear in most AI books:

The tool you choose doesn't matter nearly as much as the way you use it.

You could implement the most powerful AI platform on the market — and still fail to improve member engagement, decision-making, or retention — if you haven't aligned it to a clear strategy.

Because tools don't create clarity.
Strategy does.

When your team knows *why* they're doing something, *who* it serves, and *how* you'll know it's working — that's when change sticks.

That's when confidence grows.
That's when resistance turns into excitement.

I've seen it again and again — a team that's burned out on "yet another initiative" suddenly finds new energy when they can connect the dots between their purpose, their priorities, and the tech that supports both.

What Makes SMART Different?

SMART works because it was born inside the association world — not adapted from a corporate blueprint.

It was built:

- For overstretched CEOs
- For Boards who need to see ROI
- For teams with mixed digital confidence
- For members who are craving more care, more value, and more personal connection.

This is a strategy that respects the legacy of associations while embracing their future.

It doesn't ask you to start over.
It helps you **start smarter**.

Now that you know why SMART matters, we're going to unpack each pillar one by one.

For each, I'll show you:

- What it means in practice
- Where associations get stuck

- How AI can support you
- Real-world actions to start today.

You don't need to do it all at once.
You just need to take the first step — with strategy, not speed, leading the way.

Let's begin with the "S" in SMART — Scalable.

Because progress that overwhelms is never sustainable.
But progress that compounds? That's where the magic happens.

S is for Scalable — Start Small, Grow Smart

When association leaders hear "digital transformation," their eyes often glaze over.
Not because they don't care — but because they assume it means:

- Huge budgets
- Massive system overhauls
- Long project timelines
- Fancy consultants and complicated change processes

Let's stop that myth right now.

True transformation doesn't begin with disruption.
It begins with **direction** — and the smallest meaningful action you're willing to take today.

That's what **Scalable** means in the SMART AI Strategy™.

You don't have to launch a full member platform overhaul.
You don't need to implement five tools at once.
And you certainly don't need to wait until next financial year.

You start with *one* question:

What is one low-effort, high-impact way AI can help us serve better?

The Power of Small Experiments

The most effective AI journeys in associations don't begin with a strategy document — they begin with a *pilot.*

A lightweight, low-risk experiment to test an idea and build confidence.

Let me give you a real-world example.

An association I worked with was struggling with declining newsletter engagement. Their open rates had been dropping for years, and staff felt demoralised by the lack of response.

They didn't revamp their entire communications calendar.
They didn't invest in a new CRM.
Instead, they ran a simple AI experiment:

- They used a free subject line optimisation tool powered by machine learning.
- They A/B tested subject lines for a single newsletter issue.
- They chose the winner based on predicted engagement.

That's it.

The result?
A 27% increase in open rate — from one change.

That small win gave the team confidence. They started testing AI for content suggestions, personalisation triggers, and automated scheduling.

No massive budget.
No all-hands innovation meetings.
Just smart, scalable change.

Start With What You Already Do

You don't need to add more to your team's plate.
Instead, look at what you're already doing — and find ways to make it smarter, faster, or easier with AI.

Examples:

- If you already run webinars → use AI to analyse the chat transcript and summarise key themes
- If you're writing board reports → use AI to draft sections based on CRM or member feedback data
- If you send event reminders → use AI to tailor the messaging based on member type or past behaviour

The question isn't *what new thing can we build?*
It's *what existing thing can we optimise?*

"Scale" Doesn't Mean "Go Big"

Here's a mindset shift:
To scale well, you must start small.

Why? Because small pilots allow you to:

- Test real impact.
- Course-correct quickly
- Involve the right people.
- Build internal buy-in
- De-risk larger rollouts

Let's say you're considering an AI chatbot for member support. Instead of launching it across your whole site, test it on one high-traffic page (like "Renew Membership" or "Join Now").

Track:

- How many people engage.
- What questions they ask?
- Whether it reduces staff workload

Once you know what works, you can scale with confidence — not guesswork.

What Scaling Smart Looks Like in Practice

Let's compare two common scenarios:

Ineffective Approach	**SMART Scaling Approach**
Launches 5 tools at once	Tests 1 tool with 1 team or use case
Creates a 12-month roadmap with no delivery	Creates a 2-week sprint to try and learn
Waits for perfect alignment	Prioritises momentum over perfection
Aims to impress the Board	Aims to improve real member or staff experience
Implements tech without training	Builds staff confidence alongside rollout

Scaling isn't about expanding fast.
It's about **expanding wisely** — with clarity, evidence, and heart.

What Holds Leaders Back From Scaling Smart?

In my work with associations, I see three common blockers when it comes to scalability:

1. **Perfection Paralysis**
   Leaders wait until the tech is perfect, the budget is ideal, and the plan is airtight. But in waiting, momentum dies.
2. **Fear of Failure**
   Boards worry that a failed pilot will reflect badly on leadership. But the truth? A failed pilot is just fast learning — and much cheaper than a failed launch.
3. **Overthinking the Optics**
   Some teams want to keep pilots quiet "until they're ready." But sharing small wins early builds buy-in and makes people feel part of the progress.

If any of this sound familiar — breathe. They're normal. What matters is that you don't let them stall your first step.

LEADERSHIP REFLECTION

Ask yourself:

1. *What are we already doing that could be smarter with AI?*
2. *What's a simple pilot we could run in the next 30 days?*
3. *How will we measure success?*
4. *Who will champion it internally?*
5. *What's the smallest win that would give our team confidence?*

You don't need big tech. You need small bravery.

Bottom Line

Scalability doesn't mean *fast* — it means *fit for purpose.*
It's not about showing off shiny tools.
It's about **finding the right pace and place to begin.**

Start where you are.
Use what you have.
Do what matters.

Because when smart meets small?
That's where momentum is born.

M is for Measurable — Define What Matters, Track What Counts

Let's be honest:
Most associations track what's easy, not what matters.

Open rates. Website clicks. Event registrations. Social media likes.

Useful? Sure.
But transformational? Not really.

Being measurable isn't about drowning in dashboards.
It's about defining what success actually looks like — and making sure everyone on your team can see it, name it, and work towards it.

Because what gets measured gets managed.
And what gets managed? Gets momentum.

The Measurement Trap

One association I worked with proudly showed me their analytics platform — dozens of dashboards, hundreds of data points.

But when I asked a simple question —

"How do you know if a member feels more connected to your organisation this year than last year?" —
they paused.

They didn't have an answer. Not because they didn't care.
But because **they hadn't defined what mattered**.

That's the trap: confusing data collection with insight.
More numbers don't mean more clarity. It often leads to more confusion.

Measurability Creates Focus

Let's flip the mindset.

Imagine a new member joins your association.
You've welcomed them. They've signed up for CPD. They've joined your Facebook group.

LEADERSHIP REFLECTION

Ask yourself:

1. *What outcome do you hope to see after 90 days?*
2. *What behaviour would tell you they're engaged?*
3. *What signal would alert you if they were drifting?*

Suddenly, you're not thinking about just clicks or page views.
You're thinking about **connection, progression, retention, and trust**.

Now you're measuring what counts.

Defining "Success Signals"

Every association needs to build its own **success signals** — a set of small, meaningful indicators that show:

- A member is thriving.
- A team initiative is landing.
- A tech solution is delivering real value.

Examples of success signals might be:

- A new member logs in three times in their first month
- A webinar attendee downloads a follow-up guide.
- A support request gets resolved on first contact.
- A lapsed member re-engages after receiving a tailored nudge.
- A board receives a KPI dashboard that links engagement to retention.

None of those are vanity metrics.
They're behavioural indicators that show alignment between your purpose and your performance.

How AI Enhances Measurement

This is where AI shines — not because it generates more data, but because it helps you:

- Detect patterns.
- Spot early signals of decline or disengagement
- Generate insights based on large data sets.
- Provide real-time feedback to refine strategy.

For example:

- AI can alert you if a segment of members hasn't opened an email in 60 days.

- It can analyse webinar chat logs to uncover recurring questions.
- It can segment member behaviour to help you create better content pathways.

You're not guessing anymore.
You're designing based on real behaviour, in real time.

Stop Tracking Everything. Start Tracking the Right Things.

Let's kill the idea that **more data equals better decisions**.

Too much measurement is a productivity killer. It creates report fatigue and decision paralysis.

Here's how to fix it:

- Choose 3–5 "north star" metrics that align with your core goals
- Assign ownership — someone who is *responsible* for tracking and acting
- Review monthly in leadership meetings — not once a year
- Use AI to detect anomalies or predict trends
- Focus on *learning*, not *judging*

It's not about proving your worth.
It's about improving your impact.

Here's what it looks like when SMART becomes reality.

Case Example: Turning Data into Direction

One association I advised was struggling to retain members beyond their second year. They assumed the problem was pricing.

But AI analysis of engagement data told a different story:

- First-year members were highly active.
- In Year Two, participation dropped sharply.
- Emails weren't opened.
- Events weren't attended.
- Content wasn't accessed.

Why? Because the onboarding experience was strong — but the second-year journey was… nonexistent.

Once they saw that, they built a Year Two strategy:

- Personalised learning paths
- Invitation-only peer groups
- AI-generated check-ins after key milestones

Result?
Year-two retention jumped by 19% in one year.
All because they tracked *the right thing* — and let AI tell them the truth.

LEADERSHIP REFLECTION

Ask yourself:

1. *What are we measuring just because it's easy?*
2. *What does "success" look like — for members, for staff, for the Board?*
3. *Are we tracking leading indicators (that predict outcomes) or just lagging ones (that confirm what's already happened)?*
4. *Who owns the data? And who uses it?*

You don't need to measure everything.
You just need to measure what moves the mission forward.

The Bottom Line

Measurability is not about data — it's about direction.
It helps your team know what to focus on.
It gives your Board confidence.
It gives your AI tools something useful to work with.

Because what's the point of implementing AI if you can't tell whether it's helping?

In the next section, we'll talk about what might just be the biggest game-changer in your toolkit — **agility**.

Because in today's world, speed beats perfection.

A is for Agile — Move Fast, Adapt Faster

Let's be honest:
Most associations weren't built to move fast.

They were built for stability.
For stewardship.
For governance, risk management, and consensus decision-making.

And there's nothing wrong with that — unless it becomes a reason to resist change.

In a world where technology evolves by the week, speed has become a leadership advantage.
Not reckless speed.
Not sloppy speed.
Responsive speed. Strategic speed. Courageous speed.

That's the kind of **agility** we're talking about in the SMART AI Strategy™.

The New Definition of Agile

Being agile doesn't mean abandoning structure.
It means building systems that can **bend without breaking**.

It means:

- Testing ideas quickly
- Learning from the results
- Pivoting when needed — without ego or drama.
- Making progress *even when things are messy*.

Because here's a truth every association leader needs to hear:

Perfection is slow. Progress is fast.

Why Associations Struggle with Agility

Most of the time, it's not technology that slows you down — it's culture.

Boards want every decision justified. Teams want signoff from five people. Projects get postponed because "we're not ready."

Meanwhile, members are out there looking for answers, connection, and value.

Here are three agility blockers I see all the time:

1. **Fear of getting it wrong** – so nothing is tried at all.
2. **Approval overload** – turning simple actions into 10-step processes.
3. **Legacy habits** – "We've always done it this way" becomes a reflex.

Sound familiar?

These aren't bad people making bad decisions.
They're smart people trying to protect what they've built.

But agility asks a different question:

"What happens if we don't change?"

How AI Can Make You More Agile

Here's where AI becomes your ally.

AI tools let you:

- Test a message with 3 variations and see what lands.
- Quickly summarise survey responses into themes
- Predict what content will engage different member segments.
- Draft versions of a report or a comms piece for faster iteration

This doesn't replace human judgement — it **amplifies your speed to insight**.

It's the difference between spending two days writing a member email…
vs. testing a personalised version with AI in 20 minutes and learning what actually works.

Agility in action.

Case Example: Iterating in Real Time

One national education association wanted to run a campaign to promote an underused resource in their member portal. In the past, this would've involved:

- Copywriting
- Committee review
- Scheduling weeks ahead

This time, they tried something new.

Using a generative AI tool, they:

- Created three tailored email versions in under an hour.
- Sent to three different member types.
- Used AI to track engagement and recommend follow-ups.

Within 48 hours, they knew what worked — and pivoted accordingly.
The result? 40% increase in click-through rate compared to the original "one-size-fits-all" approach.

They didn't spend more. They moved smarter.

That's agility.
That's what's possible when we let go of "perfect" and embrace *test and learn.*

Building a Culture of Experimentation

Agility isn't just a tactic.
It's a mindset.

If your team believes they have to "get it right" the first time, innovation dies.
But if they're empowered to **test, learn, and improve**, magic happens.

Here's how to encourage it:

- Celebrate smart experiments — even when they fail.
- Shorten decision chains for low-risk actions.
- Use retrospectives (quick reviews) to learn after each pilot.
- Reward speed of learning, not just success

When people feel safe to try, they become bold.
When they become bold, your association starts to evolve faster than the landscape around it.

That's how you lead the future, not chase it.

What Agile Looks Like in Associations

Old Model	**Agile Model**
Long planning cycles	Short testing sprints
Top-down decision-making	Empowered teams
Fear of mistakes	Encouragement to try
Waiting for perfect data	Acting on 80% clarity
Post-mortems after failure	Learning loops throughout

Being agile isn't about breaking things.
It's about making things better, faster — without losing sight of your mission.

LEADERSHIP REFLECTION

Ask yourself:

1. *Where are we overthinking instead of experimenting?*
2. *What's one decision we could delegate to move faster?*
3. *Are our systems built to respond to real-time feedback?*
4. *Do we give people space to test ideas and reflect on results?*

If you wait for certainty, you'll always be behind.
But if you move with clarity and courage?
You'll become the kind of organisation that doesn't just *adapt* to change — it drives it.

The Bottom Line

Agility is your superpower in the AI age.
It lets you respond, not react.
It lets you innovate without panic.
It helps you serve in ways that feel timely, personalised, and human.

The best part?

You don't need a team of developers or a Silicon Valley budget to be agile.
You just need the **mindset of movement**.

Because speed, when aligned with purpose, creates trust.

And trust?
That's the currency of every successful association.

R is for Relevant – Put Members at the Centre of Everything

Let's get straight to it:
Relevance is your real currency.

Not reputation.
Not tradition.
Not the number of years you've existed.

Relevance is what keeps members renewing, engaging, referring, and advocating.
It's what makes your emails matter and your events meaningful.
It's what turns quiet loyalty into active participation.

In an AI-powered world, where attention is scarce and options are endless, **the only associations that thrive are those who stay deeply connected to what their members need today.**

That's what the "R" in SMART stands for.
Relevant — because if you're not relevant, you're invisible.

The Illusion of Member Satisfaction

Many associations assume they're doing fine. After all, complaints are low.
People show up to the AGM.
Your feedback surveys score well enough.

But here's the hard truth:

Silence doesn't mean satisfaction.
It often means apathy.

And apathy is far more dangerous than criticism.
When a member is dissatisfied, at least they still care enough to speak up.
When they're disengaged? They disappear quietly.
Unsubscribed. Unrenewed. Forgotten.

That's why being relevant isn't something you measure once a year.
It's something you **build into every decision, every message, every service.**

What Relevance Looks Like in Practice

Relevance is personal.
It means your member feels:

- Seen
- Heard
- Understood

- Supported

It means they look at your email and think:

"That's exactly what I needed."
It means they attend your event and say:
"I felt like that session was made for me."
It means they log into your portal and find what they were looking for in seconds — not after 14 clicks and a password reset.

And when you get it right?
It's electric.
Members don't just stay — they lean in.

How AI Helps You Stay Relevant

AI isn't about automation for its own sake.
It's about **insight-driven personalisation** at scale.

Here's what that means:

- AI can learn from member behaviour — clicks, logins, downloads — and suggest what they're likely to need next.
- It can segment your member base, so you don't send the same message to a first-year student and a 20-year veteran.
- It can draft content variations tailored to different member types.
- It can trigger personalised check-ins based on usage patterns or inactivity.

You're not guessing anymore.
You're **listening** — with data, with empathy, with purpose.

Real-World Example: A Tailored Experience

A health-focused professional body I supported had a robust CPD program, but low participation. Members told them the options "weren't relevant." The team didn't understand — the topics were valuable!

So we used AI to analyse:

- Member feedback
- Learning behaviour
- Completion data

We discovered a pattern:
Mid-career professionals were skipping offerings not because they weren't useful — but because they felt too "entry-level."

We created an AI-powered learning pathway builder:

- Members chose a focus area.
- AI recommended a three-month roadmap tailored to their goals and experience level.

Participation doubled in six months.
Satisfaction jumped.
More importantly — members felt seen.

That's what relevance looks like.
It's not about doing more.
It's about doing *what matters most.*

Don't Just Create More — Create Better

Here's a common trap:
When engagement drops, teams assume they need to create more.

More webinars.
More emails.
More content.

But quantity doesn't create relevance.
Connection does.

Ask:

- Are we solving real member problems?
- Are we making things easier, faster, more useful?
- Are we inviting members into a two-way relationship — not just pushing information at them?

AI can help you listen at scale — but *only* if you're willing to act on what you hear.

The Role of Staff and Culture

Relevance doesn't live in your CRM.
It lives in your team culture.

Do your staff:

- Understand who your members really are?
- Know what different member journeys look like?
- Feel empowered to personalise support, content, or comms?
- Celebrate curiosity and continuous learning?

Because AI can surface insights.
But only humans can act on them with care.

And relevance without humanity?
It's just creepy targeting.

LEADERSHIP REFLECTION

Ask yourself:

1. *What assumptions are we making about what members want?*
2. *When was the last time we asked open-ended questions and truly listened?*
3. *Are our programs and services designed from their perspective — or ours?*
4. *How are we using technology to deepen, not dilute, our connection?*

If you're not serving your members as they are *today* — not as they were five years ago — you're already behind.

But the good news?
Relevance is a choice — one you can make right now.

The Bottom Line

Relevance is not a one-time achievement.
It's a habit.
A mindset.
A culture.

And in the AI era, it's more achievable than ever — **if you use your tools to listen, personalise, and serve with intention**.

Because when members feel that what you do is made *for them* — they don't just stay.
They stay loyal.

They stay active.
They stay advocates.

And that's how associations grow — not just in numbers, but in meaning.

T is for Tech-Enabled — Let the Tools Do the Heavy Lifting

Let's clear something up right now:
You don't need to become a tech company.

You don't need to code.
You don't need to hire a Chief AI Officer.
You don't need to understand the mechanics of machine learning.

What you do need…
To know what's possible — and how to apply it wisely.

That's what being *tech-enabled* means in the SMART AI Strategy™.

It's not about being obsessed with the latest tools.
It's about selecting and using the right tools that make your work — and your impact — easier, clearer, and more scalable.

Technology should never distract you from your mission.
It should **free you to deliver it more powerfully.**

Tools Should Serve People — Not Replace Them

One of the biggest fears I hear from association leaders is this:

"If we bring in AI, will it replace the human touch our members value?"

The answer is simple:
Only if you use it wrong.

When done well, technology doesn't replace your people — it releases them.

It allows your:

- Membership team to focus on meaningful conversations.
- Comms team to craft better stories faster
- Events team to deliver more tailored experiences.
- Board to make decisions based on real-time data, not guesswork.

Imagine if your tools took care of the admin grind — and your people got to do the work that actually energises them.

That's the power of being tech-enabled.

What Tech-Enabled Associations Actually Do

Let's be practical.

Here's what a tech-enabled association looks like in real life:

Old Way	Tech-Enabled Way
Manual renewal reminders	Automated, behaviour-triggered nudges
One-size-fits-all email	AI-personalised messages based on role or interest
PDF-based CPD logs	Interactive, mobile-responsive dashboards

Old Way	Tech-Enabled Way
Annual member survey	Real-time feedback loops via AI text analysis
Spreadsheets for everything	Integrated systems with auto-sync

Being tech-enabled doesn't mean "digital for digital's sake." It means using *smart tools* to create *human-first* experiences — while also lightening the load on your team.

Case Study: Using AI to Reduce Admin Burnout

A national allied health association came to me overwhelmed. Staff were:

- Drowning in member enquiries
- Chasing renewals manually
- Rebuilding lost email templates every few months
- Burning out

They didn't need a rebrand.
They didn't need a complete system overhaul.

We started with one thing:
Adding an AI-powered help desk assistant trained on their FAQ and member services data.

It took 10 hours to set up.
It now handles 60% of inbound questions — instantly, 24/7.

What happened next?

- Staff stress dropped dramatically.
- Response time fell from 48 hours to under 10 minutes.
- More time freed up for relationship-building and member outreach.

Same team.
Same mission.
Now powered by tools that take care of the repetition — so the people can focus on what matters.

How to Choose the Right Tools

With so many AI tools out there, it's easy to get overwhelmed. Here's how to simplify your choices:

1. **Start with the problem, not the platform**
   → What's taking too long? What's costing you energy?
2. **Ask what could be automated or improved**
   → Is this something a machine could do faster, more accurately, or more consistently?
3. **Look for tools that play well with others**
   → Can it integrate with your CRM, comms stack, or event system?
4. **Prioritise ease of use**
   → If your staff won't use it, it doesn't matter how powerful it is.
5. **Pilot before committing**
   → Test it with one team or task before rolling out wider.

You're not looking for the "best" tech.
You're looking for the **best fit for your people, process, and purpose**.

What Being Tech-Enabled Isn't

Let's be clear about what this doesn't mean:

- It doesn't mean chasing every new app
- It doesn't mean replacing jobs with robots
- It doesn't mean losing your warmth, care, or community
- It doesn't mean spending tens of thousands on overbuilt systems

Being tech-enabled is about **sustainable, meaningful transformation** — not digital chaos.

You're not digitising your association.
You're **amplifying its humanity with better tools**.

LEADERSHIP REFLECTION

Ask yourself:

1. *What tools are our team using right now — and which ones are creating more friction than freedom?*
2. *Where are we still relying on manual effort that could be automated or supported by AI?*
3. *Are we choosing tools based on features — or based on fit?*
4. *How can we make technology feel like a partner, not a pressure?*

This isn't about getting tech "perfect."
It's about getting tech *aligned.*

The Bottom Line

To be tech-enabled is to be **future-fit**.
Not flashy. Not overbuilt.
Just fit — for your purpose, your people, and your pace.

Your mission deserves tools that lift it higher.
Your team deserves systems that support their energy.
Your members deserve experiences that feel made *just for them.*

Let the tools do the heavy lifting.
So your humans can do the heavy loving — the real, relational, purposeful work only they can do.

The Transformation

Twelve months later, the results were clear:

- Member retention rose by 15%.
- Event attendance increased 33%.
- Staff burnout dropped.
- The Board was asking, "What's next?" instead of "What now?"

And the CEO?
She didn't just feel more tech-literate — she felt more *in control* of the future.

That's what the SMART AI Strategy™ delivers.
Not overnight change.
Not hype-driven panic.
Just calm, clear, human-first progress — step by step.

SMART Reflection Prompt

Your AI Readiness in Real Life

You've now explored the full SMART AI Strategy™. Before you turn the page, take a moment to pause and reflect:

SMART Pillar	**Ask Yourself…**
Scalable	What's one small, no-risk AI pilot we could launch in the next 30 days?
Measurable	Are we tracking what truly matters — or just what's easiest to measure?
Agile	Where are we overplanning instead of testing?

SMART Pillar	Ask Yourself…
Relevant	When was the last time we truly listened to what our members need *right now*?
Tech-Enabled	What tools are draining energy instead of saving it — and what could replace them?

You don't have to have all the answers. You just need a willingness to start small and stay curious.

Next Step: Share your reflections with your leadership team or Board. Use this prompt as a conversation starter — not just to plan, but to act.

You don't need to do everything. You just need to begin — and build momentum that lasts.

Let's move into the next chapter, where we'll explore how to take your association's first (or next) confident step toward AI adoption — with focus, heart, and purpose.

Momentum in Action: The SMART AI Strategy™ at Work: How One Association Used All Five Pillars to Transform Their Future

To bring this all together, let me walk you through a composite example based on the many associations I've worked with — the kind of story you'll likely recognise because parts of it may reflect where you are right now.

This mid-sized national association had been around for 30 years. Proud history. Committed team. Deeply respected in their industry. But in recent years, they'd started to feel the pinch:

- Member renewals were down.
- Event attendance was flatlining.

- Staff were overwhelmed.
- The Board was nervous but cautious.
- Their digital tools were outdated and underused.

They weren't failing — but they weren't thriving either.

And the CEO knew they couldn't keep doing what they'd always done. She didn't want a revolution. She wanted progress. So she reached out to explore the SMART AI Strategy™ as a framework for rethinking how they worked.

S: Scalable — They Started Small

Rather than redesigning everything at once, they chose one simple pilot:

Use AI to personalise subject lines for their next three newsletters.

It took two hours to set up.
Result? Open rates increased by 28%. Engagement jumped. Staff excitement returned.

That tiny win became the spark that lit the fire. The CEO didn't push a full AI overhaul. She asked, "What else could we make smarter?"

M: Measurable — They Redefined Success

Instead of drowning in data points, the leadership team agreed on five key metrics aligned to their strategic goals — from onboarding engagement to second-year retention.

They used AI to analyse membership drop-off points and found something surprising: members were disengaging *not* because of content quality, but because of communication timing. They were sending emails too late, too infrequently, and too generically.

By defining these new success signals and tracking them weekly, they were able to act faster and more intentionally than ever before.

A: Agile — They Tested Fast, Learned Faster

The CEO introduced a "48-hour test" culture. No more six-week committee waits for every idea.

Want to try an AI-powered event description generator?
Try it this week. Review feedback on Friday. Keep what works.

They also created a "fail forward" board slide, showcasing pilot projects — even those that didn't go as planned. This normalised experimentation and gave staff permission to learn in real time.

R: Relevant — They Reconnected With Member Needs

AI analysis of website behaviour and event registrations revealed an insight: a growing segment of members were shifting careers and needed a different kind of support.

They launched a mini campaign targeting these members with a curated, AI-personalised email series. It included transition stories, reskilling resources, and a virtual networking event.

Feedback was overwhelming: "This made me feel seen."
Retention among that segment jumped 22%.

T: Tech-Enabled — They Let the Tools Lift the Load

By month six, they'd added three key AI tools:

- A chatbot answering member FAQs 24/7
- An event content recommender based on past attendance.

- A dashboard for real-time feedback on digital resource usage

The magic wasn't just in the tech — it was in how they used it.

The team no longer felt overwhelmed.
Board reports were faster and richer.
Member emails were smarter and better timed.
And most importantly, the staff had more time for meaningful connection.

PART 2

Unlocking Systems That Scale

"Simplicity isn't just about doing less. It's about clearing away distractions so you can do more of what truly matters."
– *Annie Gibbins*

CHAPTER FOUR

FROM STRATEGY TO MOMENTUM

Why Most AI Initiatives Stall (and How to Avoid It)

Let's be honest.

Most association leaders already know they need to change. They've heard the buzzwords, sat through the webinars, and maybe even tested an AI tool or two.
So why aren't they seeing momentum?

The answer isn't intelligence.
It's not intent.
It's *implementation.*

The real challenge isn't starting — it's starting *in the right way.*

Over the past few years, I've had countless conversations with CEOs, COOs, comms directors, and Board chairs who were excited about AI... but stuck in place. And almost every time, the reasons were *very human* — and very avoidable.

Let's unpack the most common reasons why well-meaning associations stall when trying to bring AI into their operations — and how we can sidestep them with confidence.

1. They Start Too Big

The most common trap? Trying to do too much, too fast.

A full tech stack overhaul.
An AI-powered chatbot before you've even cleaned your CRM.
An "innovation taskforce" with no roadmap and no remit.

These projects often sound exciting at first — until they collapse under their own complexity.

- Big bang = big burnout.

The solution?
Start smaller than feels comfortable.
The goal isn't a flashy launch. The goal is traction. Proof. Confidence.

Start with one pilot:

- Automate one member email series.
- Use AI to summarise survey results.
- Run a split test on two pieces of event copy.

Small is smart. Small is scalable.
Small gets you out of "talking" and into "testing."

2. They Don't Know Where to Begin

"I get the potential — but I don't know what to prioritise."
I hear this all the time. And it's fair.

There are hundreds of tools out there. New ones every week. It's easy to fall into research mode and never get out. But the truth is: you don't need to be an expert in AI. You just need to know your own bottlenecks.

Start with what's already slowing you down.

- Are your team spending hours responding to repetitive member questions?
- Are you rewriting the same content again and again?
- Is your event promotion or engagement falling flat?

That's where AI can help first.

You don't need a roadmap to start walking — you need a direction and the courage to take a step.

3. They Treat It Like an IT Project

Here's the danger: handing off AI to "the tech team" or expecting your IT provider to build you a strategy.

This isn't just about software.
It's about leadership, communication, and *culture*.

If the executive team isn't engaged, AI becomes a disconnected side project — and fizzles out.
If the comms team doesn't know what's possible, they won't try anything new.
If the Board only hears about risks, not benefits, they'll veto before you can show results.

This isn't about delegating. It's about **co-leading**.

You don't need to build the tools yourself.
But you do need to champion the why, model the mindset, and help your team see this as an enabler — not another burden.

4. They Wait for a Budget That Never Comes

Let's talk about money.

Yes, some AI solutions cost money. But most of the most effective tools I see associations using today are either free or under $100/month — or even included in software you already pay for (like your CRM or email marketing platform).

Here's the truth:

- You don't have a budget problem. You have a *focus* problem.

If you can save 20 hours of staff time per month with a $40 tool, that's not a cost — it's a win.

AI doesn't have to be a line item. It can be a *line lifter* — removing friction from what's already costing you too much time, energy, and opportunity.

5. They're Waiting for Permission

This is the quiet one — but I see it everywhere.

The comms manager who wants to experiment but thinks the CEO won't approve.
The CEO who's curious but doesn't want to spook the Board.
The Board who's open but assumes no one's ready.

Meanwhile… nothing happens.

The change you're waiting for. It's probably waiting for you.

You don't need a full strategy document to start. You need a test. A learning loop. A small, clear win that shows what's possible.

And once that win is visible?
Buy-in gets a whole lot easier.

So What *Does* Create Momentum?

Here's what I've seen work again and again:

- One leader who's willing to go first
- One problem that's already causing friction
- One tool or test that's low-cost and low-risk
- One result that sparks internal curiosity
- One conversation that turns into a shared goal

That's it. That's the start.

And once you get those first 1–2 wins under your belt, the narrative shifts from "Why AI?" to "Where else could we use this?"

The Bottom Line

If you've felt like you're falling behind, I want you to hear this clearly:

You're not late. You're early — if you start now.

This isn't about being perfect.
It's about being purposeful.

It's not about impressing the market.
It's about serving your members better, faster, smarter.

And it starts with letting go of the idea that AI transformation has to be overwhelming.

You don't need to leap.
You just need to **step — clearly, simply, and confidently.**

Creating Your AI Momentum Map™

Big shifts don't need big blueprints.

What they need is a **clear sense of where you stand now**, where the friction is, and where a small smart step will unlock bigger progress.

That's why I created the **AI Momentum Map™** — a simple tool designed to help association leaders move from "I don't know where to start" to "I know exactly where to focus first."

This isn't about long planning sessions or expensive consultants.
It's about clarity. Simplicity. Forward motion.

Let's walk through how to build your own AI Momentum Map™ — and use it to move from idea to impact.

What Is the AI Momentum Map™?

It's a practical framework that helps you:

- Identify where your association has capacity and confidence.
- Pinpoint where friction or overwhelm is stalling action.
- Spot quick wins that build trust and engagement
- Choose the *right-sized* place to begin — without stress.

Think of it as a compass, not a blueprint.
It won't map every detail of the journey.
But it will help you take the next step — and that's where all real momentum begins.

How to Build It (In 30 Minutes or Less)

You don't need to overthink this.
Grab your leadership team or just sit down with a pen and a cup of tea.

Here's what you do.

Step 1: Review the 5 SMART Pillars

On a whiteboard or document, list the five pillars:

1. **Scalable** — Can we experiment in small, low-risk ways?
2. **Measurable** — Do we know what success looks like and how we'd track it?
3. **Agile** — Are we open to learning quickly and adapting as we go?
4. **Relevant** — Are we closely aligned to current member needs?
5. **Tech-Enabled** — Do we have tools (or access to tools) that could support this?

Next to each one, give yourself a simple rating:

- Green: Confident — we're strong here
- Amber: Neutral — we're inconsistent or unclear
- Red: Stuck — we need support or change here

It doesn't need to be perfect — it just needs to be *honest*.

Step 2: Identify Your Leverage Point

Now, look for **the pillar that's already green** — or at least amber — and ask:

"What's the *simplest* thing we could do here to create a visible win?"

That's your starting point.

It could be:

- Testing subject line optimisation (Scalable)
- Reviewing member retention data to find trends (Measurable)
- Rewriting one piece of content with an AI co-pilot (Agile)
- Surveying members with a one-question pulse check (Relevant)
- Automating an event confirmation email (Tech-Enabled)

This is the opposite of a five-year digital transformation plan. This is a **30-day progress sprint** — a signal to yourself and your team that momentum is possible.

Step 3: Choose a Champion

Every successful shift I've seen has one common factor:

Someone owns the energy.

This doesn't need to be a new job title or another committee.

It just needs to be one person who says:

"I'll take this first test, run it, and report back."

Often, it's a curious comms lead, a proactive CEO, or a team member who's already playing with AI on their own time. Whoever it is, give them permission to learn, iterate, and move without fear of getting it wrong.

Because the only way to build momentum is to *let it move.*

A Real-World Example

One mid-sized association I worked with created their AI Momentum Map in a one-hour strategy workshop.

They quickly saw:

- They were strong on Relevance (knew their members well)
- Weak on Tech-Enabled (old systems, no clear owner)
- Neutral on Measurability (reporting was manual and inconsistent)

So they started with Relevance.

A staff member ran a two-week pilot:

Using AI to rewrite three key member emails based on different member personas.

They A/B tested versions and measured open rates and click-throughs.

One small win.
Increased engagement.
And suddenly, the Board was asking:

"Could we use this to personalise our onboarding experience?"

That's momentum.

What to Avoid

Let's quickly cover some traps I've seen associations fall into:

- **Waiting for everything to be green.** (It never is.)
- **Trying to fix red zones first.** (Start with traction, not trouble.)

- **Turning the Momentum Map into a full strategy doc.** (It's meant to spark action, not analysis.)
- **Doing it alone.** (Even 2–3 team members can spark far better insights than a solo leader.)

Your Turn: Build It Now

Before you move into the next section, take 30 minutes with your team (or yourself) and answer these five questions:

1. Which SMART pillar are we strongest in today?
2. Where are we feeling the most stuck or unclear?
3. What's one small, visible experiment we could try this month?
4. Who will lead the test and track the result?
5. What would success look like — and how would we share it?

That's it. You've just built your first AI Momentum Map™.

The Bottom Line

You don't need to solve everything.
You just need **clarity on what matters now**.

The AI Momentum Map gives you that clarity.
And from clarity comes action.
And from action? Confidence.

Because the real secret of smart leadership in an AI world isn't having all the answers.
It's being brave enough to go first — and wise enough to know where to start.

Gaining Buy-In Without Burnout

Introducing AI into an association isn't just about tools.
It's about trust.

Trust from your team.
Trust from your board.
And trust from your members — that you're evolving *with them*, not ahead of or against them.

The number one reason AI pilots stall isn't technology. It's people.
And it usually happens when leaders skip one vital step: **meaningful buy-in.**

Let's explore how to build support and enthusiasm without overloading anyone — or triggering the kind of fear that makes change feel like a threat instead of a gift.

Why People Resist — and How to Shift It

When people push back on AI, it rarely sounds like "I don't want to change."

It sounds like:

- "We're already flat out."
- "This feels risky."
- "We tried something like this years ago, and it didn't work."
- "This isn't in my role."
- "What if this replaces me?"

These aren't excuses. They're *signals.*
Signals of anxiety, uncertainty, and exhaustion — often layered over genuine care.

Remember: People don't resist change. They resist *painful* change.

The good news? You don't need to force it.
You need to reframe it.

1. Anchor AI in Purpose, Not Pressure

AI isn't a buzzword.
It's a way to **serve your purpose more effectively**.

Start your conversations here:

"This isn't about adding more to our plates — it's about taking things *off* them."
"We're not replacing people. We're releasing their capacity."
"This helps us deliver *more value* to members with *less effort*."

Leaders who start with tools lose their team.
Leaders who start with *why* build buy-in.

2. Make the First Win Theirs

If the first AI win serves the executive team, the board may disengage.
If it helps the board but burdens the staff, resentment grows.

But when the *first visible win* improves life for the people doing the work every day?

Now you're talking traction.

Start with:

- Reducing repetitive tasks
- Speeding up newsletter creation
- Answering member FAQs
- Streamlining content repurposing
- Auto-generating meeting summaries

Let your team feel the difference — in their stress levels, time, and outcomes.

When people feel helped, they lean in.
When they feel replaced, they check out.

3. Use Language That Calms, Not Confuses

Let's be real. AI terminology can feel like alphabet soup:

GPTs, LLMs, NLP, APIs, automation flows…

For most people, it's intimidating.
And when people feel out of their depth, they disengage or resist.

So how do you fix it?

De-jargon the journey.

Instead of:

- "We're using a generative large language model."

Try:

- "This tool helps us draft things faster — and we always review before it goes out."

Instead of:

- "Let's implement conversational AI in our member portal."

Try:

- "Let's test a tool that can answer common member questions instantly."

Clarity is confidence.
Confidence is momentum.

4. Let Staff Be the Innovators

You don't need to lead every AI experiment.

In fact, your momentum will multiply when you let your team play, test, and share.

Try this:

- Run an "AI Play Week" where each team member trials one tool.
- Ask them to answer two questions:
    1. What did this save me? (Time, effort, stress?)
    2. What would we need to do to use it more widely?

Now you're building *internal ambassadors*, not just compliance.

5. Earn Board Buy-In with Data, Not Drama

Boards don't need hype.
They need clarity.

Here's what works:

- Share the *cost of not changing* (time wasted, missed engagement, manual error)
- Share a real member comment from a pilot test.
- Share how your competitors are moving.
- Share a low-risk 30-day test with clear boundaries and success metrics.

And always bring it back to the strategy:

"We're not chasing AI — we're staying aligned with member expectations and future-proofing how we deliver our mission."

Boards respond to stewardship, not slogans.

6. Invite Members Into the Journey

AI isn't something you need to hide from your members. In fact, involving them early can *build loyalty*.

Try this:

- Let members vote on which new feature or improvement you try next.
- Share behind-the-scenes content on how you're testing AI to better serve them.
- Ask members what's *not working* in your comms — and show how you're using tech to improve it.

When members feel consulted, not just served, you build a relationship — not just a transaction.

What Buy-In Looks Like in Practice

A client association of mine was struggling to gain traction. The CEO believed in AI but felt like she was dragging the board and staff behind her.

So we changed tactics.

- Let the events team test an AI tool to rewrite event bios.
- Shared one pilot result with the board: 43% increase in click-throughs.
- Had one staffer do a 5-minute demo at the next team meeting.
- Sent members a short survey with one question: "What do you wish we made easier for you?"

Everything changed.

The team felt involved.
The board felt informed.
The members felt heard.

That's real buy-in.
That's how momentum grows.

The Bottom Line

You don't need to convince everyone.
You need to include them.

Lead with empathy.
Frame with clarity.
Start with service.

Buy-in isn't about getting everyone on the bus.
It's about showing them where it's going — and why the ride will be better for them, too.

In the next section, we'll look at how to **build internal capability** so that AI isn't a one-off project — but a long-term advantage that grows with your team, not around them.

Building Internal AI Capability — Without Hiring a Data Scientist

Let's be real. You're probably not going to hire a machine learning engineer or a data scientist tomorrow. And you don't need to.

The myth that AI adoption requires a team of Silicon Valley tech wizards is one of the biggest blockers in the not-for-profit and association world. It's also totally false.

AI capability isn't about having experts.
It's about building **confidence**, **curiosity**, and **consistency** inside your team.

In this section, we'll explore how you can build the internal capability to adopt AI sustainably — with the people and resources you already have.

1. Start With Curiosity, Not Competency

You don't need a team of coders.
You need a team of *learners.*

The best AI-ready teams I've worked with had one thing in common:
They asked great questions.

- "Could this tool help us do that faster?"
- "What's something we do every week that could be automated?"
- "What's frustrating our members right now — and how can we improve that?"

You don't need all the answers.
You need a culture where people feel safe enough to ask new ones.

This is where leadership sets the tone.
If you as CEO or ED can say:

"We're going to explore this together. We'll start small. We'll learn out loud."

That creates psychological safety.
And that's where real capability begins to grow.

2. Use What You Already Have

Most associations already have tools that can incorporate AI — they're just not using them to their full potential.

Here's what that looks like:

- Your email platform already includes **AI-powered send-time optimisation.**
- Your CMS probably has **content suggestions or readability analysis.**
- Google Docs now uses **AI to rewrite text or summarise meeting notes.**
- Even Zoom and Teams offer **automated transcripts, summaries, and action items**

The capability is *already there* — you just haven't turned it on yet.

Start with a quick audit:

- What tools are we already using?
- Do they have AI features we're ignoring?
- What's one we could try this month?

You may be far more "AI-ready" than you think.

3. Train the Team (Without a Conference Budget)

You don't need a full-day workshop to get started.
In fact, I recommend starting with **micro-learning**.

Here's what that can look like:

- **15-minute lunch demos**: "Let's show how this tool helped us save 30 mins on content creation."
- **Peer-led skill shares**: "Each month, one team member shares a new tool they tested."
- **AI Fridays**: One hour per fortnight dedicated to playing with a new tool and reporting back.

Want to go a step further?

- Curate a short list of YouTube videos, LinkedIn posts, or case studies tailored to your team's needs.
- Bookmark ChatGPT prompt libraries that relate to your sector.
- Invite a guest expert to run a 45-minute online Q&A — low cost, high value.

The point isn't perfection.
It's *practice.*

4. Build a Culture of Shared Learning

Capability doesn't stick unless people are given permission to keep learning out loud.

So what does that look like?

- Encouraging people to bring an AI-powered draft, *not* a polished product
- Recognising experiments, even if they "fail"
- Creating space to reflect: What did we learn? What's next?

Try this team ritual:

Every Friday, ask:
"What's one thing we learned this week about working smarter?"

Make it safe. Make it simple. Make it consistent.

Because if your team feels like they're expected to *get it right* straight away, they'll freeze.
If they know they're supported to *figure it out*, they'll fly.

5. Document What Works (and What Doesn't)

Want to build capability faster than your competitors?

Start creating a living AI playbook — written by your team, for your team.

Include:

- Tools you've tested.
- Prompts that worked well.
- Use cases that saved time.
- Mistakes you won't repeat.
- Lessons learned.

You don't need it to be fancy.
Just start a shared doc. Add as you go.

This builds institutional memory — so knowledge doesn't walk out the door when someone changes roles.

It also makes future onboarding easier, smoother, and smarter.

A Real-Life Example

One association I worked with didn't have a tech lead.
They had three staff, a WordPress site, and a Mailchimp account.

But they started an "AI Test Hour" every Tuesday.

In month one:

- They learned how to use ChatGPT to rewrite course descriptions.
- Used AI to suggest subject lines for their next email blast.

- Found a free Canva plugin that auto-resized and branded their social posts.

By month three:

- Their email open rates were up 21%.
- Their team was spending *less* time on admin and *more* time on member engagement.
- They'd created a shared doc titled "Cool Stuff We Tried That Worked."

No new hires. No budget blowouts.
Just capability built on curiosity and consistency.

The Bottom Line

You don't need a five-year roadmap.
You don't need a new hire.
You don't need a tech degree.

You need to give your people the space, safety, and structure to explore what's possible — one step at a time.

AI isn't about replacing your team.
It's about **amplifying** them.

The more confident they become, the more creative, collaborative, and impactful your association will be.

Scaling What Works — From Pilot to Practice

You've started small.
You've tested something that worked.
You've built buy-in and grown capability inside your team.

Now comes the real opportunity:

Turn momentum into a system.

This is the part many associations miss.
They launch one successful AI pilot — and then move on.
No process. No reflection. No next step.

But if you want lasting change, you can't stop at "that one project that worked."

You need to *build a feedback loop that scales.*

Let's explore how to turn your early AI wins into repeatable practices that help your association grow — not just today, but long-term.

1. Debrief What Worked (and Why)

After a successful test or pilot, most teams do one of two things:

1. Celebrate quietly and move on
2. Turn it into a formal case study that no one reads.

Neither one helps you scale.

Here's what to do instead:

- Set aside 30–60 minutes with the people involved.
- Ask five questions:
    1. What problem did we solve?
    2. What tool or tactic made the biggest difference?
    3. What did we learn that surprised us?
    4. What would we change next time?
    5. Where else could we apply this approach?

Capture the insights.
Keep it in plain language.
Store it where others can find it.

This is how knowledge becomes culture.

2. Systemise the Simple Stuff

Once you've had a win, don't reinvent it — **repeat it.**

Ask:

- Can we templatise this?
- Can we automate it?
- Can we train someone else to do it?

For example:

- If an AI tool helped you write better event descriptions, create a prompt template.
- If summarising surveys saved hours, build it into your quarterly workflow.
- If using AI for email testing improved engagement, set it as a recurring task.

Scaling isn't about doing more.
It's about making what works *repeatable* and *easy*.

Create a checklist.
Make a mini-process.
Name the task.
Assign ownership.

This is how great ideas become standard practice.

3. Share the Win Across the Organisation

Success breeds success — but only if people *know about it.*

Don't let your pilot become a quiet success in one team.
Spread it like wildfire.

Here's how:

- Send a short "What Worked" email to the team or board.
- Present the pilot outcome at your next staff or strategy meeting.
- Ask the team member who led the test to demo what they did.
- Add a "Team AI Win of the Month" to your internal comms.

The point isn't ego — it's energy.

When people see their peers succeeding with small changes, their resistance lowers and their willingness to try grows.

4. Track and Build on the Right Metrics

Scaling also means understanding what matters.

Too often, organisations track:

- Clicks, not conversations.
- Volume, not value
- Tasks done, not time saved.

You want to track what actually improves outcomes:

- Member satisfaction
- Staff capacity and wellbeing
- Retention and renewal rates
- Speed to deliver value.
- Cost or time reduction from new processes

These are the metrics that tell the real story — not of "how busy we are" but of **how well we're doing what matters most.**

Keep it simple.
Keep it visible.
And keep reviewing it as you scale.

5. Use Success to Justify the Next Step

Success earns you something precious: **permission to grow.**

Now that you've proven something works, you can:

- Request small budget increases.
- Allocate more team time for exploration.
- Ask for governance support to embed AI into operations.
- Build a roadmap that expands AI across member experience, events, advocacy, and internal comms.

This is your pivot from *project* to *practice.*
From "that cool thing we tried" to "this is how we work now."

And when done right, it never feels forced.
Because your people already feel the benefits.

A Quick Checklist: From Pilot to Practice

Here's a mini checklist to help you scale intentionally:

- We've documented what worked
- We've built a simple, repeatable process
- We've shared the win across teams
- We're tracking impact with meaningful metrics
- We've identified where to test next
- We've given someone ownership of scaling

If you can check 4 out of 6 — you're on the right track.
If you've hit all 6 — your association isn't just adapting. It's *evolving.*

From Pilot to Practice in the Real World

Here's how it looked in practice for a mid-sized membership association.

They started by automating a simple welcome email using an AI draft assistant. The result? A 47% increase in new member engagement — and team time saved. From there, the events manager tested AI-generated bios. The comms team used AI to help draft one blog per week. The CEO shared the results in a board briefing, and suddenly, everyone leaned in.

In six months, they'd gone from cautious curiosity to confident integration — without hiring anyone new or blowing the budget. The tools weren't the secret. The shift came from testing, sharing, and building one small success at a time.

The Bottom Line

AI transformation isn't a moment.
It's a mindset — and a method.

It doesn't come from one big change.
It comes from a series of small, meaningful shifts that grow stronger with each cycle.

Test → Learn → Repeat → Expand → Share → Scale.

You don't need a flashy innovation team or a million-dollar grant.
You need a way to build what works — *into the way you work.*

That's how AI becomes a strength, not a stress.
That's how your association future-proofs its purpose.
That's how your team moves from busy… to bold.

Where Is Your Momentum Now?

In the next chapter, we go beyond theory. You'll discover how associations like yours are turning small pilots into lasting progress, one confident decision at a time. From tech hesitation to team buy-in, we'll map out the real-world steps to move forward with clarity, momentum, and heart.

CHAPTER FIVE

THE CLARITY GAP

Here's the truth most leaders don't admit:

It's not that we don't want to act.

It's that we don't know where to begin — or if the next step we take will actually matter.

After the inspiration of a keynote, the energy of a strategy day, or the buzz of a new idea… reality hits.

You're back at your desk.
The emails are waiting.
The meetings are scheduled.
The budget is tight.
And the team?
Busy. Cautious. Maybe even a little sceptical.

That's where most association leaders stall — **not from a lack of desire, but from a lack of clarity.**

Insight Isn't the Problem — It's the Starting Point

Let's get something clear: if you've read this far, you *already* know more than enough to make meaningful change.

You've seen:

- Where associations are losing momentum
- How AI can unlock capacity and relevance
- What buy-in, culture, and scaling look like in practice?

The insight is there.

The question is:
What do you do next?

Because clarity isn't just knowing what's possible — it's knowing *what's next for you.*

Why the Next Step Feels So Big

Let's talk about why the "next step" feels so hard — especially for busy leaders like you.

You're not just managing tasks. You're managing risk.
You're not just thinking about this week. You're thinking about funding cycles, stakeholder trust, strategic plans, and member retention.

Taking the wrong step can feel expensive.

So what do we do?

We freeze. We research more. We wait for perfect timing. We tell ourselves:

"Let's revisit this after EOFY / the AGM / the conference / the new hire…"

But if every decision has to wait until the stars align, momentum dies.
And your association stays stuck in good intentions and great ideas — with no traction.

The Antidote to Inertia: Decide Smaller

When we say, "start small," we don't just mean pick something easy.
We mean **get closer to the work.**

What's one friction point you could ease in the next 14 days?

- Can you test an AI summariser for your next board agenda?
- Can you use AI to draft your next newsletter?
- Can you spend 10 minutes rewriting a member renewal email with a better tone of voice?

You don't need to fix everything.
You just need to take one clear, confident step — and build from there.

Action breeds clarity.
Clarity builds confidence.
Confidence leads to momentum.

Don't Overestimate Complexity — or Underestimate Impact

One of the biggest mistakes I see association leaders make is assuming that because AI is powerful, the solutions must be complex.

But in practice, the smallest changes often create the biggest shifts:

- Auto-tagging member emails to prioritise support.
- Using AI to create tailored event descriptions for different audiences
- Helping your team reduce admin through summarised meeting notes.

- Streamlining comms between departments with shared prompt templates

These aren't sexy.
They won't make headlines.
But they *will* unlock time, energy, and value — fast.

And the moment your team sees that?
They'll want more.

Don't Just Plan the Step — Own It

One more thing.

Ownership creates movement.

When no one owns the next step, it fades into "someday."
When *someone* owns it — even informally — it becomes real.

So be honest:

- What's your next best step?
- Who owns it?
- What does success look like?
- When will you review it?

You don't need a spreadsheet. You need a stake in the ground.

Let's Name It

To close this section, I want you to write something down — even if it's just for yourself.

Answer this:

"The next thing we'll try, test, or shift is…"

That's it.

That's the difference between a good idea... and a real outcome.

Choosing the Right First (or Next) Use Case

By now, you've hopefully accepted this truth:

You don't need a perfect plan.
You need a relevant one.

That means choosing your first (or next) AI application not based on trends, but based on:

- What matters most right now in your association?
- Where your team feels the most friction
- What's realistic to test with minimal risk and maximum insight.

Let's break that down — because getting these right changes everything.

The Biggest Mistake: Chasing the Coolest Tool

If your first AI use case is based on what's shiny, new, or trending on LinkedIn... you'll lose momentum quickly.

Why?

Because your team is busy. Your resources are finite. Your members care about outcomes — not innovation theatre.

So before you even open a new AI platform, start with this question:

"What's one part of our work that feels harder than it should?"

You're not chasing tools. You're solving problems.

4 Smart Criteria for a First AI Use Case

Whether you're just starting or expanding your AI use, look for use cases that meet at least **3 of the 4 criteria** below:

1. Painful

It solves something frustrating — a manual, repetitive, error-prone, or time-sucking task that no one enjoys.

Examples:

- Writing the same kind of copy from scratch over and over
- Rebuilding agendas or event blurbs for every meeting
- Responding to FAQs in email or social DMs

2. Visible

Others in your organisation (or your members) can clearly see the benefit. That visibility builds buy-in.

Examples:

- Members get clearer comms.
- The board sees better reports.
- Your staff feels a lift in workload.

3. Low-Risk

It doesn't break anything if it goes wrong. You can test, learn, and tweak without damaging trust.

Examples:

- Drafting content that still gets reviewed by humans.
- Automating internal summaries
- Tagging or categorising (not sending) outbound emails

4. Repeatable

It's not a one-off task. You can do it weekly or monthly, so it becomes a habit — not a one-hit wonder.

Examples:

- Monthly reports
- Weekly social content
- Recurring event promo copy

When a use case hits all four? That's gold.

Sample Use Cases by Department

Here's a short list of real-world use cases I've seen associations implement with success — sorted by function:

Area	Use Case
Comms & Marketing	AI-generated social post ideas based on event themes
	Optimised email subject lines and call-to-actions
	Quick rewrites for newsletters in different tones
Membership	AI-based segmentation for personalising renewal emails
	FAQs turned into chatbot-style support
Events	Session blurbs rewritten for SEO and engagement

Area	**Use Case**
	Bios and speaker intros crafted in minutes
Operations/Admin	Meeting transcripts and action item summaries
	Quick-start agenda templates with AI suggestions
Board Reporting	Quarterly impact reports with data visualisation tools
	First drafts of strategy updates or reflections

You don't have to choose more than one. Just start where your current pain is highest, and visibility is easiest.

A Real Association Example

One mid-sized member organisation was stuck. They knew AI was on their radar, but everything felt too "big."

We asked one question:

"What's something you do every single month that feels annoying — but necessary?"

The answer: their monthly CPD calendar.

It took two staff almost two days to pull together titles, descriptions, links, and emails. So we ran a one-month test using AI to:

- Generate descriptions based on bullet points.

- Suggest better titles.
- Draft two versions of the member promo email.

In the first month, the team saved 9 hours.
The open rate jumped by 11%.
The board noticed the results.
The team felt proud.

That one use case led to three more within two months — because they picked the right starting point.

Not Sure Where to Start? Ask Your Team These Questions:

Here's a quick team discussion prompt to help identify a great starting point:

- What's something you're doing manually that we repeat often?
- Where do we lose the most time (or energy) each week?
- What could we try fixing in 30 minutes or less using an AI tool?
- Where would an improvement help *you* do your job better?

These conversations surface the invisible pain points — and unlock high-impact, low-risk experiments.

Don't Start with What's Easy. Start with What Matters.

Look — you can absolutely play with AI just for fun. But if you're serious about creating internal belief, you need to pick something that makes a real difference.

Your first use case should be small — but significant.

Make sure your team **feels** the benefit.
Make sure your members **notice** the lift.
Make sure your board **understands** the value.

Then… repeat it. Share it. Scale it.

Deciding What's Worth Doing — The AI Filter for Associations

You're now in a powerful — and slightly dangerous — place.

You've seen what's possible.
You've got team interest.
You've probably bookmarked a dozen tools and templates.

But here's the trap:

Momentum without filtering leads to chaos.

Not every idea deserves your time.
Not every tool is right for your association.
Not every use case is worth testing — even if it looks exciting.

That's why this section exists. To help you **decide what's actually worth doing** — and what can wait.

You Don't Need More Ideas. You Need a Smarter Filter.

Too many associations get distracted by:

- One enthusiastic staff member pushing a shiny tool.
- A consultant promising magic with a paid platform
- An executive seeing a demo and saying, "Let's do that too!"

But if you try to chase every idea, you'll stretch your team thin — and kill confidence fast.

That's why the most effective leaders aren't just AI-positive. They're **AI-discerning.**

They ask:

- Does this actually move the needle?
- Can we afford the cost — in attention, time, or trust?
- Is this solving a real problem, or just creating busy work?

That's what the **AI Filter** helps you answer.

Introducing: The 5-Part AI Filter for Associations™

When someone brings you a new idea — or you have one yourself — use these five questions to pressure-test it before you commit.

1. Purpose

Does this align with our mission and strategy?

If the use case doesn't strengthen your ability to serve members, build relevance, or achieve your goals — it's a distraction.

Ask:

- How does this idea connect to what we care most about?
- Will it help us deliver on our promise or just add noise?

Example: Using AI to personalise member onboarding = aligned
- Using AI to create a staff karaoke night playlist = maybe not

2. People

Who benefits — and who carries the load?

Always ask:

- Who does this actually help?
- Whose job will be impacted?
- Is there clear ownership, or are we assigning this to "everyone"?

You need to be sure that:

- Someone *wants* the outcome.
- Someone *owns* the process.
- Someone *has capacity* to test it.

If the benefit is vague and the ownership is unclear, press pause.

3. Pain

Is this solving a real problem?

Sometimes, we throw AI at things that aren't actually broken — just because we can.

You want to focus on use cases that ease a real burden.

Ask:

- Is this replacing a manual, repetitive, or high-friction task?
- Will this make work meaningfully better, faster, or easier?
- Is this solving something our team or members have *complained about?*

No pain = no urgency = low impact.

4. Proof

Can we test this quickly — and learn fast?

Before you commit time or budget, ask:

- Can we trial this in 1–2 weeks?
- Can we measure the result?
- Will we know quickly if it's worth continuing?

If you can't test it quickly, it's too early.

This is key to momentum — test, learn, decide.

5. Posture

Are we ready to take responsibility?

This one's often overlooked.

It's about cultural readiness — not just technical capability.

Ask:

- Are we open to trying something that might not work the first time?
- Are we willing to share learnings across the team?
- Do we have the mindset to adapt — not just adopt?

If your team still sees AI as a threat or gimmick, the best tech in the world won't stick.
You need posture first — openness, curiosity, and clarity.

Run New Ideas Through the Filter

Let's imagine your events team says:

"We want to use AI to generate speaker questions ahead of time."

Here's what your filter might look like:

Filter	Yes / No
Purpose	Yes — improves session quality and attendee engagement
People	Yes — events team benefits, clear ownership
Pain	Yes — currently takes hours to prepare questions manually
Proof	Yes — can test with one upcoming speaker
Posture	Yes — team is open, already tested AI for session blurbs

This is a great use case to pursue.

But if the answers had been "Maybe" or "Not sure" across the board?
You'd table the idea and move on.

One Leader's Approach

A CEO I recently worked with took this filter and turned it into a weekly rhythm.

Every Monday, her leadership team would spend 15 minutes reviewing:

- Any new AI ideas from staff
- One idea from the previous week that had been tested.
- One use case that met the filter and was ready to scale.

They weren't just adopting tech.
They were **building a system of discernment**.

That's how you stay focused, grounded, and aligned.

Your Turn: Run the Filter on One Idea

Choose one idea — a tool, use case, or experiment you've been thinking about.

Answer these five questions:

1. Does this support our purpose?
2. Who benefits — and who's responsible?
3. What pain does this solve?
4. Can we test it quickly?
5. Are we ready to try, learn, and share?

If you've got at least four yeses — go for it.
If not? Park it. Revisit later. Stay focused.

Resourcing AI Without a Budget Blowout

Let's just say it outright:

You don't need a six-figure innovation fund to start using AI effectively.

Sure, a larger budget might help you scale. But to *begin, you need* focus, creativity, and a mindset shift about where your resources already are.

Here's the good news: most associations already have what they need to begin.
They just don't realise it.

Let's examine how to resource AI in a smart, lean, and immediately actionable way.

The Real Resource Problem Isn't Budget — It's Bandwidth

If your team is like most, your biggest constraint isn't money.

It's:

- Time
- Capacity
- Confidence
- Clear permission to experiment.

That's why the first thing to "resource" isn't a tool — it's *space to test something small.*

Because even the best AI platform won't help if your staff don't have time to explore it or feel too overwhelmed to try.

So before you look for funding, ask:

- Can we carve out 1 hour a week for experimentation?
- Can we protect time to debrief what we learn?
- Can we reduce one other task to make room for one new idea?

Start there. That's your real resourcing foundation.

Low-Cost, High-Value Tools to Start With

Many of the tools that work brilliantly for associations are either free, low-cost, or offer generous nonprofit plans.

Here are a few categories and examples:

Content + Comms

- **ChatGPT**: Drafts, rewrites, tone shifting.
- **Grammarly**: AI-powered editing and tone suggestions
- **Jasper**: For content workflows (paid, but team-friendly)

Operations

- **Fireflies / Otter.ai**: Meeting transcription and summaries
- **Notion AI**: Brainstorming and documenting ideas with structure
- **ClickUp AI / Asana AI**: For teams using project management tools.

Design

- **Canva** (Pro): Content resizing, smart layout, and image generation
- **Krisp / Descript**: Audio clean-up and voice-to-text.

Don't try to explore all of them.
Pick one tool that solves a current problem — and test it well.

Borrowing Time — Instead of Buying It

One smart way associations are resourcing AI is by **borrowing time** from existing meetings or admin routines.

Here's what that looks like:

- Turning a monthly team meeting into an "AI Hour"
- Replacing 1 admin-heavy task with a 30-minute AI experiment
- Assigning one staff member 1–2 hours a week to lead low-risk trials

It's not about adding time. It's about *redeploying* it with intention.

You can even pilot this as a 4-week sprint:

- Week 1: Choose a use case.
- Week 2: Try the tool.
- Week 3: Share the result.
- Week 4: Decide to scale or shelve.

This kind of mini sprint builds muscle without burning resources.

Build Your "AI Buddy System"

No one wants to be the only person trying something new.

That's why associations that succeed with AI create *peer-based learning* — not just top-down directives.

Try this:

- Pair two staff members to test the same tool together.
- Create a shared "Prompt Library" as people explore tools.
- Invite staff to a monthly "Show + Tell" on what they've tried.

You don't need a formal AI team.

What you need is *permission to play* — and a place to share what's working.

Funding Doesn't Have to Come from IT

One of the fastest ways to stop AI in its tracks is to assume it's only the IT team's job — or only possible if it fits in the tech budget.

That's not how modern AI works.

In fact, most early use cases live in:

- Communications
- Events
- Member engagement
- Education and CPD

So if you're looking for budget or support, reframe the ask:

- Could this reduce cost or workload in comms?
- Could this improve CPD access or delivery?
- Could this free up time to serve members better?

When you frame AI as a capacity booster — not a cost centre — you open the door for better conversations.

One CEO's Approach to Micro-Funding

A CEO I worked with had no digital innovation budget. But she created a $1,000 "AI Catalyst Pool" — just enough to cover three team members to test small paid tools over 3 months.

Her team used:

- One AI survey tool for events
- One design tool to improve CPD slides.
- One AI-based video captioning tool

That pool created three internal wins, freed up hours of staff time, and gave her board the confidence to approve a $10K digital innovation line item in the next financial year.

You don't need big money.
You need early proof.

The "One Hour, One Tool" Challenge

If you're wondering where to begin, try this.

The Challenge
Choose one task your team already does often (emails, reports, bios).
Give one staff member one hour to test one tool to make it faster or better.
Then share the result.

That's it.

Repeat next month with someone else.

Small effort. Big signals. Real change.

In the final section of this chapter, we'll explore how to **communicate your early AI wins with confidence** — so your board, staff, and members see you not just as a user of AI, but as a *leader* of smart, mission-driven innovation.

Communicating AI Wins to Build Belief

Here's a truth many association leaders forget:

The best innovations don't sell themselves.
People follow what they understand, trust, and believe in.

If you've piloted an AI use case — even something small — the next step isn't to immediately scale it.

It's to **talk about it** in a way that:

- Connects the dots to your purpose
- Highlights the human benefit.
- Builds confidence (not confusion)

This section will show you how.

Why Communication is the Hidden Growth Lever

AI can do a lot. But it can't:

- Ease scepticism on its own.
- Inspire trust in the unknown
- Create a culture of experimentation without leadership.

That's your job.

And the good news is, you don't need a big internal comms team or a snappy campaign.
You just need to make small wins visible — and meaningful.

What to Share (and What to Skip)

When communicating early AI success, focus on three things:

1. The Problem

What were you trying to solve?

Keep this human and relatable.
Example: "Writing monthly CPD blurbs used to take 5 hours and three rounds of edits."

2. The Shift

What changed?

Describe the new approach simply.
Example: "We used an AI tool to generate first drafts based on dot points."

3. The Outcome

What was the tangible benefit?

This can be time saved, quality improved, energy regained, or member feedback.
Example: "Saved 3 hours, faster signoff, and positive comments from members about the clarity."

That's all you need.
Problem → Shift → Outcome.

Avoid:

- Naming tools unless it adds clarity.
- Overloading with jargon
- Making it about "tech" — keep the focus on people and purpose.

Who to Tell (and How Often)

Your goal isn't to hype — it's to build trust and normalise the journey.

Start with:

- **Your Board**

  Use short updates in CEO reports. Frame it as aligned with strategy, risk-aware, and staff-led.

- **Your Staff**

Use team huddles or internal newsletters. Celebrate wins. Invite curiosity.

- **Your Members**

    Share impact, not process. Example: "We've improved how we build CPD to make it faster and more relevant to you."

You don't need to flood them with detail.
You just need to consistently reinforce one idea:

"This is helping us serve better."

A Real-World Example

One association tested AI to summarise monthly board papers — just to speed up CEO briefings.

It worked.

So the CEO included this in the next board pack:

"In February, we trialled an AI summariser to reduce paper length without losing clarity. Directors now receive a two-page summary in plain English alongside the full board pack. The result: faster reading, more focused discussions, and fewer late-night emails."

The board was impressed.
The staff felt seen.
It cost nothing — but built belief.

That's leadership communication.

Turn Every Win Into a Story Worth Sharing

Try this next time you finish an AI pilot:

Create a 3-Slide Story:

1. **The Friction**

    "We were spending X time doing Y…"

2. **The Experiment**

    "We tested Z to see if it could help."

3. **The Result**

    "Here's what we learned — and what we'll do next."

Then ask:

- Should we share this with the board?
- Can we include this in the team update?
- Would our members appreciate seeing the result?

Simple, honest, and story-driven — that's how you build AI momentum people can get behind.

Real-World Example
One mid-tier association had no tech team, no innovation lead, and no AI line item in the budget. But their CEO saw an opportunity. She ran a simple "One Hour, One Tool" test: a staff member used ChatGPT to rewrite a dry CPD email into three tones. The team voted on the best one — and member clicks increased by 17%.

The result? The team felt empowered. The board saw value. A second test followed the next month. Within three months, AI became part of weekly workflow — not because of funding, but because someone took one small, smart step and made the result visible.

Build a Culture of Sharing (Not Showing Off)

Make space for "here's what worked" *and* "here's what didn't."

In fact, sharing failures or false starts is just as powerful — because it gives others permission to try.

Consider adding a regular rhythm like:

- "AI Wins + Lessons" in your internal newsletter
- A quarterly team roundtable
- A whiteboard with staff-submitted tips, prompts, or "aha!" moments

Your goal?
Make AI feel like part of the culture — not a program, project, or passing trend.

What Happens When You Do This Well?

When you normalise talking about AI clearly and positively, you:

- Create **psychological safety** for your team to try.
- Build **strategic trust** with your board.
- Reassure **members** that change is thoughtful and mission-aligned.
- Signal to partners and funders that you're serious — but grounded.

You become not just an implementer of AI…
But a *leader of change.*

That's the shift that lasts.

If you're feeling overwhelmed, remember these three truths:

- **Start small, but significant**
  Choose one task that's painful, visible, and easy to test. Let your first win speak louder than a dozen discussions.
- **Use a smart filter to stay focused**
  The 5-part AI Filter (Purpose, People, Pain, Proof, Posture) will keep your team aligned — and your efforts strategic.
- **Make your wins visible**
  Momentum grows when people understand the "why" and feel part of the change. Share stories. Celebrate progress. Build belief.

You don't need to chase every shiny tool. You just need to champion what works — and keep testing, sharing, and building.

In the next chapter, we'll shift from ideas into *impact tracking.* You'll discover how to measure progress — not just in tools and tactics, but in culture, capacity, and confidence.

Because what matters most isn't how much AI you use. It's how well you use it to serve the mission you care about most.

"AI won't replace human relationships; it will strengthen our capacity to nurture them" – Annie Gibbins

CHAPTER SIX

MEASURE WHAT MATTERS

Your Association's AI Readiness Scorecard

Let's be honest — it's easy to *say* we want to innovate.

It's harder to prove it.

In most associations, the conversation around AI still floats in the realm of curiosity and good intention:

- "We should be looking into this."
- "There's something here we could use."
- "Once we finish this big project, we'll have time."

But if you've reached this point in the book, you're not looking for *ideas*.
You're looking for traction.

That starts with something bold and simple:

You must measure where you are — so you can lead with focus.

Because AI doesn't just test your tech stack.
It reveals the maturity of your leadership, systems, mindset, and culture.

And here's the truth most association boards and CEOs need to hear:

It's not about how many tools you're using.
It's about how ready you are to make meaningful, measurable progress — without losing your soul.

Readiness Is More Than a Buzzword

Lots of associations say things like:

- "We're aware of the trends."
- "We're having internal conversations."
- "Our team is exploring options."

But awareness is not the same as alignment.
Interest is not the same as implementation.
And tinkering is not transformation.

Readiness is the bridge between good intentions and real results.

That's why this chapter exists: to give you a structured way to evaluate how ready your association is to *lead*, *build*, and *scale* with AI — in a way that protects what matters most.

What You Can Expect From This Chapter

This isn't about getting a gold star.

It's about getting a *clear-eyed snapshot* of where your organisation stands today — across the five areas that determine how AI will either:

- Add value.
- Add noise.
- Or quietly stall and disappear.

You'll discover:

- The five core dimensions of AI readiness.
- A simple scoring system (no tech jargon required)
- What each level of readiness means — and what to do next

You'll also get access to a digital version in the appendix, where you can assess your team, board, or entire organisation.

Because you can't lead what you don't understand.
And you can't build what you haven't evaluated.

This Is About More Than Technology

Let's be very clear:
You don't need to have a chatbot on your homepage to score well on this.

You don't need to launch a full-scale digital transformation strategy.

You just need to show that your association is willing to:

- Think differently.
- Collaborate across silos.
- Learn in public.
- Adapt with purpose.
- Measure what matters.

This isn't about becoming "tech-first."
It's about becoming **future-fit.**

The Five Dimensions of AI Readiness

The associations that succeed with AI don't just have access to tools.

They have the systems, culture, and clarity to make smart decisions — even when the landscape is evolving fast.

That's why we're not measuring "tech adoption."
We're measuring *real-world readiness* — across the five dimensions that matter most:

1. **Leadership Alignment**
2. **Operating Systems**
3. **Member Experience**
4. **Data & Insight**
5. **Culture & Capability**

Let's explore each one — what it is, why it matters, and what strong performance looks like.

1. Leadership Alignment

"You can't move fast when your leaders are stuck in the past."

AI doesn't need perfect agreement — but it does need **shared understanding** and **executive sponsorship**.

This dimension is about whether your board and leadership team:

- Understand the potential and the risk of AI.
- Discuss it as part of strategic conversations.
- Have a unified language and vision.
- Support test-and-learn experiments without waiting for 100% certainty.

LEADERSHIP REFLECTION

Ask yourself:

1. *Do your leaders understand AI beyond the headlines?*
2. *Is innovation encouraged — or quietly feared?*
3. *Are decision-makers open to change before a crisis forces it?*

Strong associations have *top-down trust* and *bottom-up traction.* They don't wait for permission — they build alignment from the start.

2. Operating Systems

"You can't automate chaos."

Even the best AI tools won't help if your systems are broken. This dimension is about how **clear, connected, and scalable** your internal operations are.

It includes:

- Documented workflows (not just oral history)
- Streamlined, non-redundant processes.
- Software that speaks to each other
- Smart automation of basic admin tasks

LEADERSHIP REFLECTION

Ask yourself:

1. *Are there manual processes we still haven't addressed?*
2. *Do staff enter the same data twice across systems?*

3. *Are everyday tasks intuitive — or workarounds?*

If your operating systems are messy, AI will only magnify the mess.
If they're clean, AI can elevate them with speed, clarity, and consistency.

3. Member Experience

"AI is not here to replace human connection. It's here to deepen it."

This readiness dimension looks at how well you:

- Understand your member journey.
- Deliver personalised experiences.
- Use feedback to shape services.
- Anticipate needs instead of reacting to complaints.

It's not about fancy interfaces. It's about empathy at scale.

LEADERSHIP REFLECTION

Ask yourself:

1. *Do you know what your members care about today?*
2. *Are communications tailored or templated?*
3. *Do you measure satisfaction — or assume loyalty?*

AI is most powerful when it enhances the parts of the experience that matter most to your members. But first, you have to know what those are.

4. Data & Insight

"Data is the fuel. AI is the engine. Insight is the destination."

You can't make smart decisions without smart data.
This pillar is about how clean, current, and accessible your data really is.

It includes:

- Having a central source of truth (not 15 spreadsheets)
- De-duplicated, up-to-date member records
- Actionable analytics (not just monthly reports no one reads)
- A culture of *using* data — not just collecting it

LEADERSHIP REFLECTION

Ask yourself:

1. *Do we trust our data?*
2. *Do we use it to make decisions — or just to tick boxes?*
3. *Can we spot patterns and trends at a glance?*

If your data isn't actionable, AI will just sit on top of guesswork. But when your insights are strong, AI can turn them into *leverage.*

5. Culture & Capability

"If your team isn't ready, your tech won't matter."

This final pillar might be the most important.
Because the success of AI isn't about software — it's about **psychological safety**, **permission to test**, and **belief in progress**.

This is about whether your team:

- Feels safe trying new tools.

- Has basic training and digital confidence?
- Is encouraged to ask questions.
- Has the capacity to explore, test, and reflect?

LEADERSHIP REFLECTION

Ask yourself:

1. *Are we rewarding experimentation — or punishing imperfection?*
2. *Do people fear being replaced by tech?*
3. *Is there time carved out to learn and share?*

A culture of fear slows everything down.
But a culture of curiosity turns your people into pioneers.

Scoring Readiness:

After you review each dimension, you'll score your organisation on a simple 0–20 scale (details in the next section). You can do it solo, or with your exec team, board, or entire staff.

The goal isn't to get a perfect score.
It's to **create clarity — and then build momentum.**

Because when you know where you stand, you know how to move forward.

The AI Readiness Scorecard

You've explored the five dimensions of AI readiness. Now, it's time to measure how future-fit your association really is — using a simple, self-assessed scorecard.

This tool isn't about passing or failing.
It's about surfacing **truth**, **traction points**, and **next steps**.

Because what gets measured gets momentum.

How the Scorecard Works

Each of the five readiness dimensions will be scored from **0 to 20**, giving you a total score out of 100.

You can complete it:

- Solo (as a CEO, board member, or staff leader)
- With your leadership team
- Or anonymously across your organisation for a broader view

Each section contains **4 prompts**. For each, rate your association as:

- **0–1: Rarely true**
- **2–3: Sometimes true**
- **4: Often true**
- **5: Consistently true**

You'll then total your scores and refer to the interpretation table at the end.

Leadership Alignment (0–20)

Rate the following:

1. Our board and leadership discuss AI with curiosity, not fear.
2. AI is linked to our strategic planning and goals.
3. Leaders support pilot projects without needing full certainty first.
4. There is a shared language and understanding of AI's potential across leadership.

Score: __ / 20

Operating Systems (0–20)

Rate the following:

1. Our core processes are clearly documented and visible.
2. We've automated at least one repetitive admin task in the past 6 months.
3. Systems talk to each other — staff rarely duplicate work.
4. Our tech stack feels like a help, not a hindrance.

Score: __ / 20

Member Experience (0–20)

Rate the following:

1. We understand what our members want — not just historically, but now.
2. Communications feel personalised and relevant.
3. We actively seek feedback and use it to shape decisions.
4. Member-facing tools are mobile-friendly and intuitive.

Score: __ / 20

Data & Insight (0–20)

Rate the following:

1. We have a central, accurate source of member data.
2. Reports are turned into insights — and actions.
3. We track engagement patterns, not just headcounts.
4. Data is shared across departments, not siloed.

Score: __ / 20

Culture & Capability (0–20)

Rate the following:

1. Staff feel safe asking questions and experimenting with new tools.
2. People are rewarded for impact, not just busyness.
3. We've offered at least one AI-related learning or discussion opportunity.
4. Our team sees AI as a partner, not a threat.

Score: __ / 20

Your Total Readiness Score: __ / 100

What Your Score Means

Score Range	Interpretation	Your Focus Area
80–100	Future-Fit	Scale what's working with strategy and visibility. You're ready to lead.
60–79	AI-Aware	Build alignment, clean systems, and share early wins to build confidence.
40–59	At Risk	Focus on culture, clarity, and quick wins. You need small experiments that prove value fast.
Below 40	Urgent Shift Needed	Start with mindset. Simplify first. Your future relevance depends on your next 90 days.

What to Do Next

Once you've scored:

- Highlight your strongest and weakest dimensions.
- Identify ONE action you could take in each to improve.
- Use the free digital version in the Appendix (or companion site) to track progress over time.

Because readiness isn't static — it grows with leadership, intention, and visibility.

This is where AI becomes more than a possibility.
It becomes a *path* — and you get to walk it, your way.

From Insight to Action: Leading Forward From Your Score

Knowing your readiness score is powerful — but what you do with that knowledge is where leadership lives.

Some leaders use results as a wake-up call.
Others use them to confirm what they suspected.
The best ones? They use it as a blueprint for where to lead next — with clarity, calm, and conviction.

Here's how to use your score in five meaningful ways.

1. Find Your Leverage Points

Don't try to fix everything. That's not strategy — that's overwhelm.

Instead, look across your five scores and ask:

- Where is the **lowest score** dragging down our readiness?
- Where is the **highest score** creating momentum, we can build on?

Your **lowest-scoring area** is your friction point. It's where change is most needed — and most likely to be resisted.

Your **highest-scoring area** is your credibility anchor. It's what your team already trusts and believes in. Leverage this to build momentum.

Insight: One meaningful shift in your lowest-scoring area can unlock progress across all others.

2. Choose Your First Win Wisely

Don't start with the flashiest tool or the biggest risk.
Start with something visible, simple, and painful enough that *everyone* wants it fixed.

Ask:

- What's an everyday friction that AI could solve in 1–2 hours?
- Who on the team would be excited to test it?
- How can we make the result visible across the organisation?

Your goal isn't just to use AI — it's to build belief.
And belief is built through small, smart wins that *matter*.

3. Tell the Story of Progress

AI culture isn't created by tools. It's created by stories.

Every time you test something new — and it works — tell the story:

- What problem were you solving?
- What tool did you try?
- What did it change (time saved, engagement lifted, insight gained)?
- What did you learn — and what's next?

Use email. Use team meetings. Use board reports.

Leaders don't just lead the work. They **narrate the momentum.**

4. Revisit Your Score Quarterly

Your AI readiness isn't a one-time exercise.

It's a mirror — and the more often you look, the more confident your leadership becomes.

Make it a habit:

- Add the scorecard to your quarterly strategy check-ins.
- Ask teams to re-score themselves as projects evolve.
- Use trends (not just scores) to shape your AI roadmap.

Insight: The goal isn't to chase a "perfect 100."
It's to **increase clarity and reduce friction — one decision at a time.**

5. Connect Score to Strategy

Finally, take your readiness results and connect them to what matters most:

- Your mission
- Your growth plan
- Your member experience
- Your operational goals

Every line in your strategic plan can be **supported or accelerated** by smarter systems, better data, clearer culture — and yes, AI.

But only if you connect the dots.

This is where AI becomes less about automation — and more about amplification.

Real-World Glimpse:

A medium-sized professional association scored 58 on their first AI readiness check — especially low in Data & Insight and Culture & Capability.

Instead of panicking, they ran one small AI test: using ChatGPT to rework a member renewal script. That alone boosted email opens by 23% — and saved the team 10 hours.

After sharing the result in an all-staff meeting, two other teams volunteered to run pilots. Within six months, they improved their readiness score to 74 — and built a shared language of innovation.

What changed? Not the tech stack.
The leadership mindset. The visibility of progress. The culture.

Final Reflection: Ready to Lead What's Next

So here we are.

You've explored the shifts in mindset.
You've unpacked the SMART AI Strategy™.
You've chosen your filters.
You've measured your readiness.
You've seen what's possible.

Now it's time to lead with that knowledge — not someday, but today.

Because here's the truth no one talks about enough:

AI isn't your destination. Your mission is.
AI is the vehicle that helps you get there with more clarity, care, and capacity.

This isn't about jumping on a tech trend.
It's about *future-proofing the heart of your association* — by reducing unnecessary effort, amplifying the human work, and freeing your team to focus on what really matters.

And you don't need a million-dollar innovation fund to begin.
You just need to do what great leaders always do:

- Ask better questions.
- Invite others into the journey.
- Build small wins into visible momentum.
- Let purpose guide your next right step.

This Is What AI-Ready Leadership Looks Like:

- You know where the friction is — and you're willing to name it.
- You see opportunity in every frustration.
- You lead culture, not just compliance.
- You use data to serve, not just to report.
- You adapt your strategy to serve real people in real time.

And most importantly? You never lose sight of your "why." Because your members don't need a robot. They need leadership. They need relevance. They need *you.*

Your Turn

Here's how to close Part One with intention:

1. **Review your readiness score.**
   Highlight one win — and one weakness.
2. **Choose one AI filter to test this month.**
   Purpose, People, Pain, Proof, or Posture. Just one.
3. **Set a 30-day mini goal.**
   Pick a task. Pick a team. Track a result. Share what you learn.
4. **Talk about it.**
   The most powerful change agents are the ones who talk *while* they build — not after.

Because the most successful associations won't be the ones with the fanciest tools.
They'll be the ones with **bold, clear, human-centred leadership.**

Let this be your moment.

The moment where uncertainty shifts to clarity.
Where hesitation shifts to traction.
Where overwhelm shifts to ownership.

You don't need to know everything.

But you do need to believe this:
Your association is not too small.
Your systems are not too old.
You are not too late.

You are right on time.

Now, let's move into Part Two — where we'll go deeper into **real-world case studies**, **use cases**, and **implementation plans** you can customise, adapt, and lead with.

Because you're not just learning AI.
You're building the future — with purpose, people, and possibility.

PART 3

Unlocking Leadership That Lasts

"Real leadership isn't about managing change; it's about inspiring confidence through uncertainty." – *Annie Gibbins*

CHAPTER SEVEN

START WHERE YOU ARE

Your First 90 Days of AI Momentum

You've laid the foundation.
You've chosen your filters.
You've measured your readiness.
Now, it's time to move from *understanding* to *momentum.*

Chapter Seven is all about **activation** — not as a grand, top-down transformation, but as a series of focused, real-world moves you can make in the first 30, 60, and 90 days to start seeing traction, clarity, and team buy-in.

Because in my experience, what separates associations that *talk* about innovation from those that *lead* it isn't the technology. It's that they act before they feel 100% ready.

This chapter will help you do the same.

The Myth of the Big Bang

Let's bust one myth right now:

AI transformation doesn't begin with a million-dollar budget or a shiny new strategy.

It begins with:

- One nagging admin problem.
- One team member willing to test something new.
- One meaningful use case that proves the value of momentum.

We're not chasing perfection here. We're building *capability* — step by step, in a way your team can understand, support, and *own*.

That's why this chapter is structured around a simple, strategic lens:

- **First 30 days**: Observation + Setup
- **Days 31–60**: Low-risk experiments + early wins
- **Days 61–90**: Consolidation + visibility + internal storytelling

You'll also see how this ties back to the SMART AI Strategy™ pillars — so each small move contributes to a larger, values-aligned shift.

But First: Anchor to Purpose (Again)

Before we dive into your roadmap, let's pause for just a moment.

Because here's the trap many leaders fall into:

- They get excited about tools.
- They trial five things at once.
- The team gets overwhelmed.
- The change fizzles.

You don't need five pilots. You need one *anchored in purpose.*

So ask:

- What member problem are we solving?
- What internal pressure are we easing?
- What cultural shift are we signalling?

If your answer doesn't feel grounded in purpose, stop.
Reframe. Reset. Then proceed.

Days 1–30 — Setup and Strategic Observation

If you're like most association leaders, you might be tempted to jump straight into action. I get it — you want results, and you want them now. But the truth is:

The fastest way to get traction is to start slow — with precision.

The first 30 days are your opportunity to pause, observe, and prepare your environment for success. You're not trying to "do more" — you're trying to remove the noise so the right things can flourish.

This phase is about three key priorities:

1. **Clarify Your AI Focus**
2. **Prepare Your People**
3. **Audit for Opportunity**

Let's explore each one.

1. Clarify Your AI Focus: Define the "Why Now?"

This isn't about picking tools. It's about choosing a purpose-aligned focus that your team can get behind.

Start by asking:

- What pain point could we remove with AI?

- What process or project is draining time or morale?
- What outcome would make people say, "That helped"?

Examples:

- You've got 400-member support emails per week? Maybe AI can help triage.
- Your newsletter open rates are falling? Try AI-generated subject line testing.
- Your events team spends 8 hours formatting post-event summaries? AI can draft those in minutes.

Pro Tip: Focus on a task your team *already hates* — not something they love. You're more likely to get support if AI is introduced as *a help*, not *a threat.*

2. Prepare Your People: The 3-Point Conversation

Before you even test a tool, get your people ready. That doesn't mean running a full training program. It means having a conversation that builds curiosity and confidence.

Here's a simple framework I've used in teams over and over:

A. Acknowledge the unknown

"We don't have all the answers — and that's okay."

B. Anchor in values

"We're not using AI to replace people — we're using it to *release* people."

C. Invite contribution

"What's one task you wish we could simplify? Let's test something small, together."

You'd be amazed how this opens the door. People don't need perfection. They need **clarity, permission, and proof you care**.

3. Audit for Opportunity: Where Is the Friction Hiding?

You can't fix what you can't see. Use your first 30 days to run what I call a "friction scan."

Ask your team to complete this sentence:

"It's ridiculous that we still do ________ manually."

You'll discover gold.

Things like:

- "Manually formatting every board agenda."
- "Writing renewal reminder emails from scratch."
- "Finding past webinar links across five folders."

These are **AI-ready moments** — because they're:

- Repetitive
- Time-draining
- Low-risk
- Currently annoying

Bonus Step:
Create a shared doc or whiteboard with all the suggestions. Let your team *see* that their pain points are heard — and solvable.

What Success Looks Like at Day 30

By the end of your first 30 days, you should have:

- One clear use case selected (small, meaningful, non-technical)

- A few internal champions who are curious and engaged
- A team that feels informed, not ambushed
- A short list of high-friction tasks AI could help with
- A mindset that says: *"This is possible."*

You haven't "done" AI yet — but you've done something even more important:
You've created **the conditions** for smart, purpose-driven progress.

Days 31–60 — Run Low-Risk Experiments and Build Credibility

You've done the thinking. You've prepared your team. Now it's time to test something real.

But here's the key:

Your first AI experiments aren't about sophistication.
They're about *signal.* The message you send. The confidence you build. The possibilities you unlock.

The goal in this stage is to run **1–2 small AI pilots** that are:

- Non-technical
- Time-saving
- Instantly visible
- Easy to reverse.

When done well, these experiments:

- Prove AI's value without disruption.
- Increase staff buy-in through experience.
- Spark new ideas across departments

Let's walk through how to do it.

Step 1: Choose Your First Pilot

Go back to your "friction list" from the last 30 days. Now pick one opportunity that:

- Involves something repetitive.
- Happens weekly or monthly.
- Isn't tied to compliance or sensitive data.
- Impacts multiple people (or members)

Here are a few real-world examples:

Use Case	Tool Used	Outcome
Auto-drafting member emails	ChatGPT / Claude	Saved 5–7 hours weekly
Creating event agendas	Notion AI	Cut admin time in half
Generating FAQs from surveys	ChatGPT	Created knowledge base in minutes
Summarising board minutes	Google Docs AI	Faster turnaround + more engagement
Writing captions for social posts	Jasper / Canva Magic Write	Boosted content consistency

The magic isn't in the tool. It's in the **relief** and **time-saving** it creates.

Step 2: Set a Learning Frame

Before you start, set expectations with your team:

- "This is a test, not a commitment."
- "We're here to learn — not to get it perfect."
- "Our goal is insight, not outcome."

Document:

- What you're testing
- What success looks like?
- What feedback you'll collect?
- How long the test will run (2–4 weeks is plenty)

This builds safety. It removes pressure. And it gives everyone permission to try without fear.

Step 3: Run the Pilot and Capture the Story

Now run it.

Don't overanalyse. Don't build a committee.
Just test, learn, adjust.

Most of these pilots can be run by one or two people using free tools. The cost is minimal. The insight? Priceless.

But here's the leadership unlock:

"People don't resist change. They resist being changed without context."
— Satya Nadella, CEO, Microsoft

Don't just run the pilot. *Tell the story.*

Share it in your team meeting:

- "Here's what we tried."

- "Here's what it changed."
- "Here's what we learned."

This isn't fluff — it's strategy.

When people see small wins shared with transparency, trust grows. Culture shifts. New volunteers step forward.

Bonus tip: Use screenshots. Time saved. Team quotes. Anything tangible.

Step 4: Debrief and Decide What's Next

At the end of the test, ask:

- Did it save time or reduce admin?
- Did people feel empowered or frustrated?
- Could we scale this without much change?
- What surprised us — in a good or bad way?

If it worked:
Make it standard.

If it didn't:
Celebrate the learning — and test something else.

Remember, **failure isn't the opposite of success in innovation.**
It's the foundation of it.

What Success Looks Like at Day 60

By now, you should have:

- 1–2 AI pilots tested
- At least one small win to celebrate.
- Documentation of results (to share with board/staff)
- Higher awareness and lower anxiety among your team

- A culture that's saying, *"Hey, what else could we try?"*

This is how credibility builds. Not in theory — in **experience**.

You've moved from talking about AI to working *with* it.
And that shift will echo through everything you do next.

Days 61–90 — Build Visibility, Share Results, and Set the Next Stage

You've completed your early wins. You've tested, learned, and made small changes.

Now, it's time to do what separates reactive organisations from strategic ones:

Turn quiet success into visible belief.
Make the learnings bigger than the pilot.
Bake innovation into how you operate — not just what you do.

This 30-day window is all about three things:

1. **Celebrate and Communicate Results**
2. **Decide What Becomes "Business as Usual"**
3. **Set Up the Next Momentum Cycle**

Let's break them down.

1. Celebrate and Communicate Results

Here's a simple rule: If no one knows what worked, it didn't happen.

Your team needs to *see and hear* the win — not just in a PDF or project folder, but in your everyday rhythms.

Use your channels:

- Staff meetings
- Member newsletters
- Slack or Teams updates.
- Board reports
- LinkedIn updates from your CEO profile

And don't just share facts — share the *before and after.*

Example:

- "We used to spend 6 hours creating these summaries. Now we spend 30 minutes — and they're better written."
- "This AI tool helped us reduce support response times by 40% in two weeks."

When people see the *impact,* they're more open to exploring what's next.

Pro tip: Give credit. Publicly thank the team or individual who led the pilot. Innovation is contagious — but only when it's rewarded.

2. Decide What Becomes "Business as Usual"

Not every test becomes a system — and that's okay.

But if something:

- Saves time.
- Improves member experience.
- Increases clarity.
- Builds staff confidence.

...it deserves to stay.

Ask:

- "What worked well enough to keep?"
- "Does it need refinement or formal process?"
- "Should we train others to use it?"
- "Where else in the org could we apply it?"

The goal here isn't just to *do AI* — it's to reduce manual effort *systematically.*

Even if you only lock in one improvement, that's success. That's momentum. That's culture change.

3. Set Up the Next Momentum Cycle

This is the secret weapon of forward-moving associations:

They don't stop after the pilot.
They create a rhythm of innovation.

Before you end your 90-day sprint, choose:

- One new area to explore.
- One friction point to solve.
- One staff member to lead a test.
- One piece of feedback from your first cycle to act on

Your second wave doesn't need to be bigger. It just needs to be **intentional.**

Ideas to consider:

- Expand an AI use case to another department.
- Pilot an AI tool for member segmentation or content personalisation.
- Test automation for social media summaries.
- Use AI to tag and analyse member survey responses.

🗝 *Leadership insight:*
You don't need to run the next pilot yourself. Empower someone else. Coach them. Share the spotlight.

That's how you go from leading innovation to **embedding** it.

What Success Looks Like at Day 90

By now, your association should:

- Have at least one AI-driven process running regularly.
- Have a story of success shared internally and/or externally.
- Be capturing ideas from across the team.
- See higher curiosity and confidence around AI.
- Be ready to start the next 90-day cycle with less friction.

This is no longer a "new thing."
It's a **way of working** — one that reduces pressure, enhances value, and amplifies purpose.

Final Reflection — 90 Days of Progress, One Step at a Time

You made it. You moved. You didn't wait for perfect conditions — you started with what you had. That's leadership.

Let's pause and take stock of what those first 90 days have actually created.

- You clarified your AI readiness.
- You chose a purpose-aligned starting point.
- You ran a low-risk experiment.
- You created and shared visible wins.
- You set the next stage with more buy-in, clarity, and control.

That's more than progress. That's **transformation, quietly underway**.

This Is the Real Power of the First 90 Days

It's not about how many tools you tested.
It's about what you *unlocked*:

✓ Curiosity
✓ Time
✓ Culture
✓ Confidence
✓ Energy

And you did it without a tech team, a huge budget, or a formal AI strategy.
You did it through intention, simplicity, and heart.

The truth? You don't need to become "tech-driven."
You just need to become **momentum-aware.**

A Quick 90-Day Reflection Checklist

Use these five questions to anchor what you've learned — and what you'll carry forward:

1. What's one process we improved or simplified using AI?
2. What's one mindset shift we noticed across the team?
3. What's one story of progress we've shared (or could share)?
4. What's one barrier we're still facing — and how could we tackle it?
5. What's one area of the association we're excited to explore next?

This isn't just about pausing to reflect. It's about recognising that leadership doesn't always show up as a bold announcement.

Sometimes it looks like a quiet win. A team nodding along. A few hours saved. A new question asked.

That's the work. That's how it builds.

Let's Be Clear: This Isn't the End. It's the Pattern.

What you've built here isn't a phase.
It's a pattern. A rhythm. A capability you can now apply again and again.

Because the real win isn't the tool.
It's the ability to ask:

- "Where's the friction?"
- "Could we try something here?"
- "What's our next best step?"

And that — more than any product or platform — is what future-proofs your association.

Case Study — What Momentum Looks Like in the Real World
One national health education association started with just a single AI pilot — automating the creation of event summaries using a free tool. In the first month, they saved eight admin hours.

By month two, they tested AI for personalising new member onboarding emails. Click-through rates jumped by 40%. That small win sparked curiosity.

By month three, they had created an "AI sandbox" team — one person from each department, meeting monthly to test small use cases.

Twelve months later, without adding headcount or tech debt, they'd:

- Reduced admin overhead by 17%
- Improved member response rates by 22%
- Introduced AI into board reporting, saving hours on data synthesis.

They didn't do everything at once. They just kept asking, "What could we improve with 10% less effort?" That mindset — not the tools — created a culture of innovation.

Use Cases in Action

You now have the confidence to lead.
In the next chapters, we'll get even more specific.

We'll explore:

- Use cases that increase member retention.
- Strategies to personalise the member journey.
- AI-supported content workflows
- Data insights without the overwhelm.
- Real association examples you can borrow and build from

Because now that you know *how to start*, it's time to look at *where to grow*.

90-Day Momentum Map: How to Lead With Confidence

Time Frame	What to Do
Days 1–30	Align to purpose. Scan for friction. Set up a pilot.
Days 31–60	Run a low-risk experiment. Capture results. Share the story.

Time Frame	What to Do
Days 61–90	Celebrate small wins. Embed success. Plan next test.

This isn't about one-off innovation. It's about creating a rhythm your team can build on.

You don't need to rush.
You just need to *repeat what works*.
Momentum is your new superpower — and you're just getting started.

“In the age of AI, data doesn’t just answer questions — it teaches us what to ask next.” — Annie Gibbins

CHAPTER EIGHT

BETTER QUESTIONS BETTER DECISIONS

How AI Turns Your Data into Direction

Let's get something straight: most associations aren't suffering from a lack of data.

They're suffering from a lack of **clarity**.

From tangled spreadsheets to scattered member feedback, to performance dashboards no one checks — data is everywhere. But insight? That's something else entirely.

You don't need more reports. You need **meaningful interpretation** — the kind that helps you make smart, strategic decisions before the moment passes.

And this is where AI becomes a game-changer.

AI doesn't just show you the numbers.
It reveals patterns, predicts behaviour, and suggests next steps — all at the speed your team actually needs.

This is your shift:

- From **data-heavy** to **data-smart**
- From **tracking everything** to **focusing on what matters**
- From **guessing** to **gaining clarity**

Let's walk through how that actually looks.

What AI Can (and Can't) Do With Your Association's Data

Let's be real — most associations don't have pristine data lakes or neatly tagged systems. What they have is much more common:

- Outdated member records
- Survey results scattered across PDFs.
- Clunky CRMs that only two staff members understand.
- Financial spreadsheets living in someone's inbox.

Sound familiar?

Here's the good news: **AI doesn't need perfect data to be powerful.**
What it needs is **accessible data** — and a clear question to answer.

Think of AI not as a magic wand, but as a magnifying glass. It shines a light on what's already there, helping you spot the patterns you've been missing.

Let's break this down into what AI actually *can* do — and where your leadership still matters most.

What AI *Can* Do with Your Existing Data

1. Clean it up — fast.
AI tools can now:

- Identify duplicate records.
- Spot missing data
- Suggest likely corrections.
- Merge similar entries (e.g. "Ann Gibbons" vs "Annie Gibbins")

Even basic integrations with platforms like HubSpot, Salesforce, or Member Jungle can use AI to improve your data hygiene without burning staff hours.

2. Make sense of scattered feedback.
Got hundreds of open-ended survey responses? Member emails? Event feedback?

AI can summarise this feedback, categorise it into themes, and even generate a list of priority issues. It gives you the *story behind the stats* — in seconds.

3. Spot behavioural patterns.
AI can analyse how your members interact with your website, emails, webinars or renewal flows. It helps answer questions like:

- "Who's most engaged — and why?"
- "Which content drives action?"
- "When do members tend to drop off?"

It's like having a digital listening post — always running in the background.

4. Forecast what's coming next.
With historical data and a little setup, AI can:

- Predict future renewal rates.
- Flag at-risk member segments.
- Estimate attendance at future events.
- Model the impact of changing your pricing.

You no longer have to rely on intuition alone. Now, you can blend gut feel with genuine foresight.

What AI *Can't* Do (and Shouldn't Be Asked To)

1. Decide what matters.
AI doesn't know your mission, your member promise, or your ethical boundaries. That's your job.

It can *inform* your decisions. But it can't *make* them for you. Especially not the hard ones.

2. Replace real leadership conversations.
Data may tell you that event attendance is dropping — but it won't explain the politics behind it, or why a key stakeholder is disengaged. You still need the human context.

3. Know what questions you're not asking.
AI is only as good as the prompts you give it. If you're not willing to challenge your assumptions, you'll just automate the status quo.

And that's not transformation. That's just digitised stagnation.

The Big Shift: Ask Better Questions, Not Just Faster Ones

What if your board stopped asking:

- "How many members joined last quarter?" …and started asking:
- "What are our top three early indicators of member dissatisfaction — and how soon can we act on them?"

What if your team stopped asking:

- "How many people opened the last email?" …and asked:

- "Which content formats build the most trust — and can AI help us scale that?"

The magic of AI is not that it gives answers.
It's that it reveals new places to look.

Better data isn't about quantity — it's about curiosity.
Better questions create better strategies. And that's where leaders shine.

How to Build a Data Culture That Works With AI — Not Against It

You don't need to become a data-driven organisation overnight. But you do need to become **data-respectful** — especially in the age of AI.

Here's what I've learned from two decades in association leadership:
Culture eats technology for breakfast.

You can buy new systems. You can hire consultants. You can even run impressive board presentations about your "data transformation roadmap."
But unless your people feel **safe**, **skilled**, and **supported** to use data, none of it will stick.

AI doesn't just need inputs — it needs **alignment**.

So let's explore how you build a **data culture** that doesn't scare your team — it **serves** them.

1. Reframe Data as Empowerment, Not Evaluation

In many associations, data feels like a trap.

It's used to:

- Monitor staff.
- Justify funding.
- Spot mistakes
- Prove or disprove ideas.

The result? Your team avoids it, underreports, or hides it.

But what if you reframed data as a mirror, not a microscope?

What if instead of asking:

- "What went wrong?" You asked:
- "What are we learning?"
- "How can we improve this without burnout?"
- "What would success look like next time?"

That's when data becomes empowering. That's when your staff lean in — because they know the numbers won't be used against them, but for them.

Leadership cue: The tone you use when asking about data sets the culture more than the metrics themselves.

2. Make Data Accessible, Not Mysterious

A common mistake? Data lives in a locked dashboard or siloed CRM — only one or two people have access, and everyone else relies on monthly exports.

That kills momentum.

When data is *centralised* and *accessible*, AI becomes useful.
When data is *hidden* or *complex*, AI becomes irrelevant.

So what can you do?

- Give team leads access to dashboards they understand.
- Use plain language to describe what's being tracked.
- Share "data snapshots" weekly (e.g. 3 metrics that matter)
- Let AI create short summaries your team can scan in 30 seconds.

You don't need to teach everyone how to run SQL queries. You just need to remove the fog.

3. Encourage Questions Over Reports

A strong data culture isn't obsessed with dashboards — it's obsessed with curiosity.

Imagine your staff meetings started with:

- "What's one surprising trend we noticed this week?"
- "Where are we seeing momentum?"
- "What feels off — and what data might help us clarify it?"

AI thrives in curious environments. It needs people willing to experiment, iterate, and learn.

Too many teams are afraid to ask questions because they assume they're "not technical enough." That's your cue to remind them:

You don't need to be a tech expert to ask smart questions. You just need permission to explore.

4. Tell Data Stories, Not Just Numbers

Raw metrics rarely inspire action. Stories do.

Let's say your event attendance rose by 12% this quarter. That's nice. But what if you said:

"By switching to AI-assisted reminders, we brought back 86 lapsed members who hadn't engaged in 18 months — and five of them signed up for mentoring."

Now that's a story people remember. And it's driven by data.

Train your team to:

- Find the narrative in the numbers.
- Tie results to member experience.
- Celebrate what worked (and what you'll test next)

This makes data feel relevant — not robotic.

5. Build AI Confidence, Not Dependence

Last but not least: AI should support your team, not replace their judgment.

Don't force adoption. Don't overwhelm with jargon. Don't make "using AI" a performance metric.

Instead, ask:

- "Where are you spending time that feels repetitive?"
- "Would it help if we tested a tool together?"
- "What's one thing you wish you had better insight into?"

Create an environment where people can try, reflect, and iterate.

That's when AI becomes cultural, not just technical.

From Hindsight to Foresight: Questions That Change the Game

One of the most powerful shifts AI brings to association leadership isn't about speed or automation. It's about **the quality of the questions we're able to ask** — and answer.

Without AI, most decisions rely on:

- What happened last quarter?
- Gut instinct
- What other associations are doing?
- A few highlighted numbers from a staff report

That's hindsight-driven leadership. And it works — until it doesn't.

AI opens the door to **foresight**: the ability to see patterns, test assumptions, and model potential futures before they arrive.

But only if you ask the right questions.

Let's walk through the five levels of association decision-making — and how AI helps you evolve each one.

Tracking → What's Happening?

This is where most associations start.

Typical questions:

- How many members renewed last month?
- What was the open rate for our last email?
- How many attendees came to our last webinar?

These are good baseline metrics. But they're rear-view. You're tracking the past.

AI can still help here:

- It can summarise reports.
- Identify anomalies.
- Visualise trends.
- Deliver results faster.

But the goal is to **move up** — from reactive to proactive.

Diagnosing → Why Is It Happening?

This is where the real insight begins.

Better questions:

- Why are first-year members not renewing?
- What content formats drive the most engagement?
- Which segments are declining in event participation — and when did that start?

AI can:

- Detect hidden correlations.
- Cross-reference data from multiple systems.
- Surface contributing factors in plain English.
- Reveal timing patterns you missed.

This is where teams start getting curious — and AI becomes a valued partner in discovery.

Predicting → What's Likely to Happen Next?

Now we're moving into foresight.
This is where AI shines — when you train it to anticipate, not just explain.

Questions to explore:

- Which members are likely to lapse next quarter?
- What will our event attendance be if we increase pricing by 10%?
- Which communication channels are losing traction?

With the right data, AI can build models that help you test scenarios — before you commit resources.

You're no longer relying on "best guesses."
You're simulating outcomes — then choosing wisely.

Prescribing → What Should We Do About It?

This is where insight meets action.

Now you're asking:

- If we want to increase early renewals, what's the best 3-step campaign?
- What's the most efficient channel for onboarding new members?
- Which benefits do lapsed members care about most — and how can we highlight them?

AI tools (especially those with machine learning) can suggest:

- Specific campaign ideas
- Content optimisations
- Member segmentation strategies
- Smart scheduling options

But again — *you* are the strategist.
AI brings options. You bring judgment, values, and vision.

Innovating → What's Possible That We Haven't Considered Yet?

This is the game-changer.

These are the questions that turn good associations into future-fit ones:

- What are the unmet needs in our profession?
- What service could we deliver better than anyone else — using AI?
- What value could we create that would make our members say, "I didn't know I needed this, but I love it"?

AI can surface patterns you weren't even looking for.

It can synthesise external trends, member behaviours, and global benchmarks.
It can even scan for emerging needs across public platforms and forums.

But again — it's your leadership that makes the leap.

Innovation doesn't come from AI.
It comes from your courage to ask the questions others haven't dared yet.

Better Questions → Better Decisions → Better Impact

So don't ask:

- "How do we keep up?" Ask:
- "What if we lead?"

Don't ask:

- "What will AI do to our association?" Ask:

- "What can AI do **with** our association — if we ask the right questions?"

Because in the end, AI doesn't replace thinking.
It enhances it.
It speeds it up.
And it opens it up to possibilities we couldn't see before.

The New Decision-Making Framework for AI-Supported Leadership
~1,000+ words

Here's the truth no one tells association leaders:

The hardest part of digital transformation isn't adopting the technology — it's making decisions with it.

That's especially true when you're suddenly surrounded by more data, faster feedback loops, and a growing pressure to "act with confidence."

But confidence doesn't come from certainty.
It comes from **clarity** — and clarity is built on better thinking.

That's where this new decision-making model comes in.

Introducing:

The CLEAR Decision Model™

Clarity. Learning. Ethics. Alignment. Results.

This isn't a tech framework.
It's a **leadership compass** — built to help association executives integrate AI-powered insights without losing their purpose, people, or priorities.

Let's break it down:

C — Clarity

What are we really trying to solve?

Start here. Before diving into the data, clarify the decision you're trying to make:

- What outcome matters most?
- Who does this impact?
- What assumptions are we bringing into the room?

AI can surface a hundred possibilities. But clarity filters out the noise.

If you can't clearly define the question, don't trust the answers.

L — Learning

🔍 *What does the data tell us — and what doesn't it?*

Ask:

- What trends are showing up?
- What's missing or underreported?
- Are we interpreting this correctly, or projecting bias?

AI can summarise insights, but you must still interrogate them. Remember: learning is iterative. AI should support a conversation, not dictate a conclusion.

E — Ethics

What are the risks, rights, and responsibilities involved?

Don't just ask what's possible. Ask what's appropriate.

Examples:

- Will this data usage respect our member privacy?
- Could this decision unintentionally exclude or disadvantage a group?
- Are we being transparent about how AI is informing outcomes?

This is where purpose-driven leadership matters most. Because just because AI *can* doesn't mean your association *should.*

A — Alignment

Does this align with our strategic goals and values?

You've got insights. You've considered the risks. Now ask:

- Will this decision move us toward our mission — or sideways?
- Are we solving a real problem or chasing a shiny object?
- Is this part of our broader AI strategy, or a one-off reaction?

AI tools should always serve your *why.*
When they drift from your core focus, they become a distraction, not a solution.

R — Results

How will we measure success — and what will we do with what we learn?

Every AI-informed decision is a test. So close the loop.

Ask:

- What does success look like — in 30 days? In 90?
- How will we share outcomes with the team and board?

- What did we learn — and how will we improve next time?

Without follow-through, even great ideas fade fast.
But with feedback, you turn decisions into data for your next round of learning.

Putting It All Together

Here's how a real-world association might use the **CLEAR Decision Model™**:

Scenario: You're deciding whether to pilot AI-powered personalised content in your member newsletter.

Clarity — We want to increase click-through rates and relevance for diverse member segments.
Learning — Our open rates are declining, especially among early-career members.
Ethics — We'll be using anonymous engagement data only — no personal profiles.
Alignment — This supports our goal of deeper member engagement and retention.
Results — We'll track open/click rates, segment performance, and member feedback over 90 days.

Simple. Smart. Strategic.

You don't need a PhD in data science to lead this well.
You just need a repeatable process — and the courage to use it consistently.

Case Study: CLEAR in Action

Let's say a mid-sized association in the environmental science sector was facing dwindling engagement from early-career members. Their instinct was to launch a new mentoring program — fast. But instead of jumping in, the leadership team used the **CLEAR Decision Model™**.

- **Clarity**: They defined the real challenge — not just a lack of engagement, but a sense of disconnection among young professional's post-pandemic.
- **Learning**: AI analysed years of survey feedback and found that what early-career members wanted most wasn't mentoring — it was career visibility and networking with industry leaders.
- **Ethics**: The team ensured any AI-generated content or recommendations wouldn't replace real human mentorship or override community norms.
- **Alignment**: They aligned their solution with their strategic priority of member-led innovation and DEI goals — testing peer-led knowledge circles instead of top-down mentoring.
- **Results**: Within three months, event participation among this group jumped by 22%, and member referrals increased for the first time in a year.

It wasn't flashy. But it was focused, inclusive, and insight-driven — powered by a better question, supported by AI, and anchored in mission.

A Culture of Smarter Questions and Shared Decisions

Here's what happens when you lead with CLEAR:

- Your board gains confidence in your use of AI
- Your staff feels empowered to explore, not afraid to fail.
- Your association stays anchored in its values — while evolving its methods

This isn't about perfect answers.
It's about principled inquiry.
That's what makes you future-fit.

Ready to Use It?

We've included a **CLEAR Decision Model worksheet** in the Appendix to help you run this process in your next leadership or strategy meeting. Use it to walk your team through tough decisions, together.

Whether it's choosing a new tech tool, launching an AI pilot, or refining your member journey — **CLEAR keeps you focused, values-aligned, and action-ready.**

In the next chapter, we'll move from leadership frameworks to lived experience — and unpack real AI use cases across associations just like yours.

Ask, Don't Assume.

AI doesn't lead for you — but it will lead you forward, if you ask the right questions.

That's the real superpower here.

Not just faster reports or cleaner spreadsheets.
But smarter prompts. Deeper patterns. Sharper foresight.
All rooted in clarity, integrity, and purpose.

The leaders who thrive in this AI-enabled era won't be the ones with the flashiest tools.
They'll be the ones who lead with questions — and invite their teams to grow alongside them.

Because when you learn to work *with* your data — not around it — you unlock more than insights.
You unlock alignment. You unlock possibility.

THINK LIKE A FUTURE-FIT LEADER

Better Questions → Better Decisions → Better Impact

Use this moment to pause and reframe:

- What are you solving?
- What are you learning?
- Are your choices aligned with your mission?
- What does success actually look like?

When in doubt, return to CLEAR — and lead from clarity, not chaos.

Your Turn

Before you move on, pause and ask:

- **What's one question your association should be asking, but isn't yet?**
- **Where are you relying on instinct, when AI could offer insight?**
- **What's one small way you could use the CLEAR model this month — to make a better decision, faster?**

Write it down. Share it with a colleague. Use it to kick off your next team conversation.

Because the smartest AI in the world still needs a human who's ready to ask better questions — and act with boldness.

"Innovation isn't about having the best technology - it's about having the courage to use it wisely." - Annie Gibbins

CHAPTER NINE

FROM PILOT TO PROGRESS

Scaling What Works

Let's get one thing clear: Pilots are not success. Pilots are potential.

If your team just finished a small AI test — a smarter email workflow, a chatbot answering FAQs, or a member data clean-up project — that's amazing. That took courage. It took effort. It probably took a few late nights and "Wait, how do I use this again?" moments.

But now the question is:

How do you turn that spark into a system?

Because unless your pilot evolves into a process — a way of working, not just a one-off — it risks becoming just another good idea that didn't stick.

The Progress Flywheel: Small Wins That Compound

Here's the truth: Smart associations don't scale with hype. They scale with rhythm. That rhythm is what I call the **Progress Flywheel**.

Think of it like this:

1. **Pilot** something small and strategic.

2. **Reflect** on what worked and what didn't.
3. **Refine** it with feedback from your team and members.
4. **Share** the early wins across your organisation.
5. **Repeat** the process — smarter and faster each time.

Small wins → Shared belief → Sustainable changetrust

The flywheel gains momentum. And that momentum builds belief.

Your board starts leaning in.
Your team gets curious instead of cautious.
Your members notice the improved experience — and stay engaged.

What Scaling Looks Like in the Real World

Let me give you a composite example from the field:

An education association started with a modest pilot: using ChatGPT to help draft customised CPD content summaries for different member tiers. It saved the education team **four hours a week** and boosted content completion rates by 22%.

Encouraged, they ran another experiment: using AI tools to analyse their survey responses for hidden trends. Within a week, they uncovered a new service need emerging among mid-career members — one they had completely missed in manual reviews.

Within six months:

- Admin load dropped by 30% in three departments.
- Event satisfaction scores rose across the board.
- Staff burnout complaints decreased.
- And most importantly? The board approved funding for a 2-year AI roadmap.

They didn't start big. They started *clever.*

From Test to Culture

Scaling isn't about tools. It's about trust.

You're not just training your staff to use AI. You're creating a culture that sees experimentation as safe, progress as possible, and success as shared.

That's when the magic happens.

When AI stops being "the innovation team's project" and becomes "how we work around here."

One of the most pivotal moments in my AI journey as a CEO was when I realised our members weren't disengaged — they were unheard. We'd spent months crafting campaigns, building programs, and running webinars, but something was missing. So, we piloted an AI sentiment tool to analyse feedback across surveys, emails, and event chats. What it showed me? People were confused. Overwhelmed. Unsure where to start. That insight changed everything. We simplified how we communicated, personalised our onboarding, and finally saw the engagement we'd been chasing. That wasn't an AI success — it was a leadership one, powered by a better way to listen.

Practical Prompts to Build Your Flywheel

To help you scale what works, ask:

- What pilot has shown early traction?
- What barriers (real or imagined) are slowing scale?
- Who can become a *scaling champion* in each team?
- What do we need to let go of to make space for what's next?

Write it down. Talk it through with your leadership team. And commit to one action that moves the wheel this quarter.

Communicating Wins Without the Hype

Here's something that often gets overlooked in association transformation:
You can't scale what your people don't believe in.

Too many AI projects fizzle out not because the results weren't strong — but because the story wasn't shared. Or worse, it was shared in a way that made people feel left out, overwhelmed, or talked down to.

The goal isn't to throw data at your board.
It's to build momentum by telling stories that matter.

Win the Mindset, Not Just the Metric

Here's a simple rule I use:

"If it worked, talk about it. If it helped people, celebrate it. If it saved time, show where that time went."

For example:

- Don't just say: "We cut processing time by 48%."
- Say: "We now have 10 more hours a week to personally follow up with new members — and satisfaction is already rising."

That's not a stat. That's a story. And it inspires belief.

The 3 Layers of a Great Transformation Story

When you're communicating your wins to your board, your staff, or your members, structure your message like this:

1. **Start with the Why:**
    - "We wanted to improve the member experience."

  - "Our goal was to reduce the admin burden on our events team."
  - "We needed better insight to make informed decisions."
2. **Share the How (Simply):**
  - "We tested a small AI tool to support that goal."
  - "We didn't automate everything — we started with one piece."
  - "We involved our staff in designing the test."
3. **End with the Human Win:**
  - "Now our members are getting faster responses."
  - "Our team feels less overwhelmed."
  - "We're making decisions in real-time — not waiting weeks for reports."

That's how you create buy-in **real people, real results, real meaning.**

Include the Sceptics — On Purpose

Want to build long-term success?

Make space for the critics — and involve them early.

That might sound counterintuitive, but here's what I've seen time and again:

The team members who are the most cautious at the beginning often become your greatest champions — *if* they feel heard and supported.

Don't dismiss resistance. Decode it.

Ask:

- "What concerns you most about this change?"
- "What would make this feel safer to try?"

- "What would be a win for you personally?"

Then act on what you hear.
Nothing scales faster than change that people helped shape.

Building a Strategic Roadmap That Isn't Just Another Plan on a Shelf

Let's be honest — most strategic plans gather dust. They sound impressive. They look good in a board pack. But they rarely live beyond the PDF.

When it comes to scaling AI in your association, you need something more dynamic, more real-time, and more *alive.*

You need a roadmap that breathes.

From Static to Strategic-in-Motion

An AI roadmap isn't a 60-page document with rigid milestones. It's a **living plan** that adapts to what works, what breaks, and what matters most.

Here's the mindset shift:

Traditional Planning: "Let's outline every step and hope reality matches."
AI-Ready Planning: "Let's map a clear direction, test early, and learn fast."

Instead of locking in multi-year deliverables, focus on:

- Clear outcomes over detailed activities
- Fast iterations over grand rollouts
- Informed experiments over fixed assumptions

The Four Anchors of a Living Roadmap

Every great AI strategy roadmap I've seen has these four elements:

1. Guiding Goals

What are the 3–5 outcomes that matter most over the next 12–18 months?

These should be member-focused and mission-aligned. Examples:

- Reduce admin friction across departments.
- Improve member renewal experience.
- Create a more responsive data culture.

2. Pilot Pathways

What small-scale projects will you run to test AI's role?

You've done some already — now group them by theme. Examples:

- Personalised comms for members
- Automation in onboarding or renewals
- AI-supported reporting for better board decisions

3. Resources + Champions

Who owns what? What support do they need?

This is where leadership makes or breaks the plan. Assign:

- Owners (not just departments)
- Budget buffers (for unexpected learning)
- Safe spaces (to fail fast without blame)

4. Feedback + Flexibility Loops

How will you know if it's working? How will you course-correct?

Design check-in points *before* you hit problems.

- Monthly team reviews
- Quarterly board summaries
- Member pulse feedback
- "Pause + pivot" discussions as needed.

Your Roadmap Doesn't Have to Be Perfect. Just Clear.

Perfectionism is the enemy of progress here.
You don't need to anticipate every turn — just mark the next mile.

Keep asking:

- What matters most right now?
- What did we learn from the last test?
- What's the next smart step?

Do that consistently — and your roadmap won't sit on a shelf.
It'll lead real change, in real time.

Managing Growth Without Losing the Heart of the Organisation

Scaling can be exhilarating — but also risky. Growth isn't just about size or speed. It's about staying *true* as you evolve.

This is where a lot of associations stumble.

They start automating.
Streamlining.
Expanding services.

Adding tools.
And somewhere along the way… they lose touch with their heartbeat.

What Makes You *You* Can't Be Automated

AI can improve what you do. But it should never replace *why* you do it.

Let's be clear:
Your members didn't join because you had cutting-edge automation.
They joined because of your purpose. Your values. Your community.

As your systems scale, your **culture must deepen.**

That means:

- Talking about your mission more often, not less
- Sharing stories of impact — not just outputs
- Creating space for human touchpoints, even when processes are digital.

You're not just growing smarter. You're growing **more human at scale.**

How to Stay Grounded as You Scale

Here are five practical ways to stay connected to your mission as your AI capabilities grow:

1. Revisit Your Purpose — Often

Make it part of decision-making conversations:

"Does this AI tool help us better serve that purpose?"

If it doesn't? Don't scale it.

2. Keep Humans in the Loop

Yes, AI can send a renewal reminder.
But a follow-up call from a real person might be what gets a member to stay.

Balance efficiency with empathy.

3. Share Member Stories, Not Just Member Stats

Showcase real voices from your community.
The success story of a student member, the feedback from a regional event, the appreciation from a long-standing committee chair.

AI can give you data.
Stories remind everyone why the data matters.

4. Protect the Rituals That Build Culture

That yearly thank-you message.
That welcome call.
That moment of recognition at your conference.

Automate the tasks so you can *elevate* the moments.

5. Create AI Guidelines That Reflect Your Values

Don't wait for an ethics breach.
Establish clear guardrails now that align with your association's tone, inclusivity standards, and member expectations.

Growth with heart isn't an accident. It's a commitment — to evolve *and* stay anchored.

Because the best use of AI isn't to move you away from your values.
It's to make it easier to *live them out at scale.*

From Scaling to Sustaining: The Power of Consistent Action

If you've made it this far — you're not dabbling anymore.
You're building.

You're no longer testing to see *if* AI fits your association.
You're now leading the charge to embed it in ways that are responsible, relevant, and real.

This is where transformation becomes sustainable.

Because here's the truth no one talks about:

It's not the boldest ideas that win.
It's the leaders who stick with them — through complexity, setbacks, and small steps forward.

Momentum beats magic.

That's how movements are made.

Reflection Prompt: What's Ready to Scale?

Pause for a moment and consider these three questions:

- What's working right now that deserves to be shared or scaled?
- Who in your organisation is quietly leading progress — and how can you support them?
- What story can you tell your board this quarter that links small AI shifts to big human wins?

If you can answer even one of those — you've already begun the next cycle of change.

Now it's time to go deeper.

Unlocking the AI Use Cases That Matter

In the next chapter, we'll move from strategy to specificity.

You'll explore the most powerful, practical **AI use cases** for associations — from member engagement and advocacy to operations, learning, marketing, and beyond.

This is where theory becomes tools.
Where experiments become ecosystems.
Where you start building an AI-enabled association — with intention, alignment, and heart.

Let's get specific.

PART 4

Unlocking Your Future

"The future isn't something we enter. It's something we create, intentionally, one strategic decision at a time."
– *Annie Gibbins*

CHAPTER TEN

FROM POSSIBILITY TO PRACTICE

Becoming the AI-Confident Association

You've Done the Work. Now Lead the Way.

If you've come this far, take a breath — and take a bow.

You've questioned the status quo.
You've explored what's possible.
You've challenged the mindset that says, "That's not for us."
You've tested, refined, communicated, and built momentum — one smart step at a time.

This is the chapter where you stop experimenting in the shadows and start stepping into the light.

Because the world doesn't need another hesitant association leader.
It needs ones who are grounded, values-led, and bold enough to **show what's possible**.

From Early Exploration to Everyday Practice

Up until now, your AI work may have lived in a pilot program, a side project, or a leadership conversation that felt "experimental."

But what if I told you this is where it starts to count?

The goal isn't to be perfect with AI.
It's to become fluent with it — familiar, not fearful.

This chapter is about turning what you've started into a shared way of working across your whole organisation. Not as an "innovation department thing." Not as a line item. But as a **leadership capability and cultural habit.**

This is where AI stops being the shiny new idea and starts being part of your organisational rhythm.

Becoming the Association Others Watch

You're no longer watching from the sidelines — you're shaping the future. And when you start using AI with confidence, clarity, and care, people notice. Your members feel it. Your peers in the sector see it. And your team stands taller because they're part of something modern, courageous, and impactful.

Being AI-confident doesn't mean you've mastered every tool. It means you've built the systems, the mindsets, and the muscle to **keep learning, keep applying, and keep leading.**

So, here's the call:

Stop hiding the great work you're already doing.
Start leading with it.

You don't need a perfect case study to speak up.
You just need a real story, rooted in service, showing what's possible when associations put purpose before perfection.

The Confidence Curve: From Curious to Credible

Let's get something clear:
You weren't expected to have all the answers at the start.

And yet here you are — with something most leaders don't have *momentum.*

That's the key difference between the curious and the confident.
Curious leaders ask, "What is AI?"
Confident leaders ask, "How do we use it to serve our people better today?"

You've moved from asking questions to creating answers.
From uncertainty to experimentation.
From ideas to action.

That progression is what I call **The Confidence Curve** — and it's one of the most important leadership journeys of our time.

The 4 Stages of the Confidence Curve

1. Curiosity

This is where most association leaders begin.
AI feels exciting but unclear — even overwhelming.
You're aware it matters, but unsure how or why.

You're reading, watching, listening, asking questions.
That's where the seeds are planted.

"Maybe this isn't just hype."
"I should probably learn what ChatGPT actually does…"

2. Capability

This is when you start applying — in small, meaningful ways.

You test a tool. You streamline a process.
You reduce admin. You personalise a message.
You bring others along for the ride.

The result? You learn by doing — not by waiting.

"That saved us four hours."
"This could work across our renewal reminders."
"Our comms team is already more productive."

3. Credibility

Now your association has proof — and people notice.

Your board is asking fewer questions and showing more support.
Your team is more willing to try new tools.
Your members are getting value, and it shows in feedback and engagement.

You're not just using AI. You're leading with it — ethically, practically, and purposefully.

"We've improved our response time and member satisfaction."
"Other associations are asking what we're doing differently."
"We're now advising our peers in the sector."

4. Culture

This is the tipping point — where AI isn't just a strategy.
It's part of who you are.

It's embedded in how your people work.
In how decisions get made.
In how innovation is approached — not as a project, but as a mindset.

This is when **confidence becomes contagious.**

"It's just how we do things here."
"We're not afraid of testing what's next."
"We trust our people to experiment responsibly."

You Don't Need to Skip Steps — You Just Need to Keep Moving

Not everyone in your organisation will be at the same point on the curve — and that's okay.

But as a leader, your role is to:

- **Model the mindset** of continuous learning.
- **Give your team permission** to try, fail, and improve.
- **Champion what's working** and share the wins.

You don't need to be the expert.
You just need to be **the enabler**.

Because confidence isn't a personality trait — it's a pattern of action.

How to Communicate Your AI Wins Without the Hype

If AI success happens and no one knows about it…, did it really happen?

One of the most overlooked parts of becoming an AI-confident association is this:
You have to talk about it.
Not with hype. Not with jargon.
But with clarity, humility, and mission-aligned storytelling.

Because when your board sees how AI is creating results, they get behind it.
When your staff sees that it's not just "more tech," they engage.
When your members feel the difference, they stay — and advocate.

Quiet Success Doesn't Scale

I've seen this firsthand:
An association runs a successful AI pilot — like automating feedback analysis or improving onboarding.
It saves time. Lifts engagement. Improves clarity.
But they never tell the story.

They move on to the next project, and all the energy… evaporates.

The team who did the work gets no recognition.
The board forgets why it matters.
The momentum fizzles out.

Your job isn't just to lead the work — it's to **make it visible**.

Not for the spotlight, but to build belief.

Three Principles of Sharing AI Wins

1. Make It About the Member, Not the Machine

Instead of saying:

"We used an NLP tool to automate feedback collection…"

Say:

"We reduced member response time by 48%, which helped us catch issues earlier and respond with more care."

People don't care about the tech.
They care about the *result* — especially when it's human-centred.

2. Turn Internal Experiments Into External Leadership

Tested something that worked?
That's not just a win — it's a story.

Post about it. Present it. Share it with other associations. It builds your credibility and contributes to the broader conversation.

Example:

"We trialled an AI-supported content assistant to improve the quality of our member resources. It now saves our team 6 hours per week — time we're reinvesting in strategic partnerships."

Keep it honest. Keep it real. Keep it focused on value.

3. Use Storytelling, Not Just Stats

Numbers tell part of the story.
But *narrative* is what people remember.

Frame each win with:

- The problem or pain point
- The small change you made.
- The real outcome for a person or team
- The learning you'd pass on.

It turns a technical shift into a *transformational moment.*

What Happens When You Communicate Well?

- Your team feels empowered and appreciated.
- Your board becomes your advocate.
- Your members start trusting you're investing in *them*, not just tech.
- Other associations look to you as a model of what's possible.

And more importantly, you build an internal culture where people say:

"Hey, let's try something — and let's share it."

That's when innovation becomes a habit, not a headline.

Positioning Yourself as a Sector Leader

The associations that shape the future aren't the ones who use the most tech.
They're the ones who use it with the most *clarity* and *care* — and then show others how to do the same.

By now, you've done more than implement AI.
You've navigated the culture. Built internal belief. Tested, refined, and scaled with purpose.

That means you're ready for the next phase:
Stepping into thought leadership.

You don't need to be the loudest.
But you *do* need to be the most grounded.
Because when your peers are still asking, "Is AI right for us?" — you've already built the playbook.

Why Association Leaders Need to Speak Up

This moment in time belongs to us.

AI in the association sector is still emerging. And the sector needs voices that represent:

- Ethics over ego
- Purpose over performance
- Service over speed

When you share what's working — and where you're still learning — you give others permission to begin. You create a ripple effect of relevance.

And make no mistake:
The sector is watching.
Your insights are more valuable than you think.

Opportunities to Lead the Conversation

You don't need a keynote or a book deal to lead.
Here's where real influence is built:

- Present at your industry's annual conference.
- Share AI wins in CEO roundtables or working groups.
- Partner with adjacent organisations to co-design ethical AI frameworks
- Write a short case study or article on your journey so far.
- Mentor a peer association leader just starting out.

And most importantly:

Be the voice that says, "This is *doable* — and worth it."

When you lead with authenticity, you're not just sharing tech strategies.
You're shaping the *character* of AI leadership in the not-for-profit and member-driven sector.

Leadership in the Age of AI Is Legacy Work

This isn't about jumping on a trend.
This is about what we leave behind.

As association leaders, we are the stewards of trust.
Of knowledge.
Of community.

If we can demonstrate how to harness technology *without losing our humanity* — we offer our sector the clearest path forward.

And when you lead like that?
You're not just future-proofing your own organisation.
You're helping to future-proof the profession.

Reflection: Are You the Leader You Needed at the Start?

Let's bring this full circle.

When you first picked up this book, you may have felt curious, cautious, or even a little overwhelmed.
You might've been wondering:

"Where do I start?"
"What if I get it wrong?"
"How can we stay relevant *and* values-led?"

Now look where you are.

You've explored the mindset, the models, the mechanics.
You've learned that AI isn't the end — it's a new kind of beginning.
One that's not about replacing people, but about *releasing* their best work.

So, pause here.

LEADERSHIP REFLECTION

Ask yourself:

1. *Am I becoming the leader I needed at the start of this journey?*
2. *Am I leading with clarity instead of control?*
3. *Am I encouraging curiosity over perfection?*

4. *Am I fostering inclusion rather than going it alone?*
5. *Am I choosing alignment over urgency?*
6. *Am I making decisions grounded in purpose, not just pressure?*

If the answer is even *partially* yes — that's powerful.
That's progress.
That's what modern leadership looks like.

And here's what else it means:

You are *already* the kind of leader who can carry your association into its next chapter — not by knowing everything, but by modelling what it means to adapt with integrity.

You've become the kind of leader others look to — not for answers, but for example.

Building a 12-Month Smart Growth Plan

By now, you've done more than scratch the surface.
You've laid a strong foundation: mindset, systems, team culture, and leadership clarity.

Now it's time to shift gears — from insight to implementation.
From momentum to a **smart, intentional growth plan**.

Because AI success isn't just about what you do next.
It's about how you keep doing what works — consistently, confidently, and with care.

In the next chapter, we'll bring all the moving parts together into a **12-month strategic framework** that aligns with your values, team capacity, and member needs.

You'll learn:

- How to set realistic, aligned goals that actually stick.
- How to create a rhythm of continuous improvement without burning out your team
- What to track, how to measure, and how to communicate progress
- How to stay responsive in a changing world without losing focus

This isn't about doing more.
It's about doing what matters — with purpose, precision, and humanity.

Get ready to design your association's **AI-enabled growth journey** — one quarter at a time.

Let's go.

CHAPTER ELEVEN

THE POWER OF PLANNING

Your 12-Month SMART AI Growth Cycle

The Danger of AI Without a Plan

You've come a long way.

You've unpacked the mindset.
You've explored the tools.
You've challenged assumptions and found new language for what leadership looks like in this new era.

But now comes the part that separates *progress* from *plateau* — and that's **structure**.

Let's be honest: most associations don't fail to innovate because they don't care or because they lack the tools.
They fail because they never put those tools into a sustainable rhythm.

They try one thing, then jump to the next.
They experiment in isolation, but never align across teams.
They get excited, then overloaded, then retreat into business-as-usual.

Sound familiar?

This is what I call the **"AI stall zone."**
It's that awkward middle place where you've started — but you haven't yet scaled.
You've dabbled, but not embedded.
You've trialled, but not tracked.

And it's dangerous.
Because that's where belief breaks down.
That's where leadership starts to wonder if this was all a waste of time.
That's where team energy drops, and momentum dies quietly.

Not because AI didn't work.
But because *you didn't work the AI into your world* — intentionally, incrementally, and with the kind of clarity that lets everyone breathe and believe at the same time.

Planning Isn't Bureaucracy — It's Alignment

Let's throw out the old assumptions.

This isn't about writing a 60-page innovation strategy that no one reads.
It's not about booking a day-long "transformation retreat" with too many post-it notes and no follow-up.

This is about having **a living rhythm** — a light, flexible, strategic guide that helps your team focus on *what matters next.*

Because when AI becomes part of your weekly conversations, your quarterly reviews, your annual roadmap — that's when it stops feeling scary and starts feeling normal.

That's when confidence builds.
That's when systems evolve.
That's when change sticks.

So, if you've ever thought:

"We've done some great stuff… but I don't know what comes next,"

This chapter is for you.

We're going to build a growth plan that works with the pace of real associations — imperfect, evolving, busy — but deeply committed to service and sustainability.

From Overwhelm to Ownership

When teams feel overwhelmed, it's rarely due to too much work—it's the weight of unclear priorities. A well-structured plan isn't about pressure. It's about *permission.*

Permission to say no to distractions.
Permission to go deep on what matters.
Permission to move forward with clarity.

I've seen teams move from burnout to energy when we replaced endless to-do lists with a focused plan. Suddenly, every effort had a purpose. Meetings got shorter. Progress became visible.

And the biggest shift? Ownership.

Instead of waiting for "leadership to decide," team members began to lead themselves. They knew what quarter they were in. They knew what mattered. They were part of the momentum, not victims of it.

That's the power of intentional planning. It doesn't stifle initiative—it sets it free.

The SMART AI Growth Framework

Before we dive into calendars and quarterly breakdowns, let's zoom out for a second.

You've been introduced to the **SMART AI Strategy™** earlier in the book — a foundational framework I created for association leaders to evaluate and elevate their approach to technology.

Now we're going to move from **strategy** to **structure**.

This section is all about applying the SMART principles — not as high-level concepts, but as practical **filters** to help you build your 12-month growth plan with purpose and clarity.

Let's revisit what SMART stands for:

- **S – Scalable**
- **M – Measurable**
- **A – Agile**
- **R – Relevant**
- **T – Tech-Enabled**

When used together, these five elements ensure your AI initiatives aren't just exciting — they're executable, sustainable, and deeply aligned with your mission.

S – Scalable

Ask: *Can this idea grow with us?*

Scalability doesn't mean launching a massive, all-at-once initiative. It means building in a way that can expand as your confidence, budget, and capacity grow.

If you implement a chatbot, can it evolve from handling three FAQs to managing full onboarding journeys over time?

If you test predictive analytics, can it be integrated across departments next quarter?

In your plan:
Start with small, controlled experiments — but choose projects that have the potential to scale over the year.

M – Measurable

Ask: *How will we know this is working?*

Measurement isn't just about numbers — it's about learning. If you're launching an AI-supported feedback tool, what metrics matter? Speed of response? Quality of insights? Member sentiment?

More importantly: who is tracking them, and how often?

In your plan:
Build in simple feedback loops and reflection points.
Set quarterly review milestones where the team can ask: *What's improved? What's stuck? What needs to change?*

A – Agile

Ask: *Can we adjust quickly if needed?*

Rigid plans fail in fast-changing environments. AI is moving fast — and your members are too. Your systems need the space to test, tweak, and pivot without fear.

Agile doesn't mean chaotic.
It means intentional **iteration** — with room to learn as you go.

In your plan:
Design 90-day action windows with mid-cycle checkpoints.
Rather than annual targets, focus on what's possible *this quarter.*
Then reassess.

R – Relevant

Ask: *Is this serving our members and our mission — right now?*

AI that doesn't align with your current challenges or your members' real needs is a distraction, not an innovation.

Relevance is about **value**. It's about knowing your audience, listening to them, and responding with precision.

In your plan:
Make relevance your North Star. Tie every AI action back to a real member need, a team pain point, or a strategic objective.

T – Tech-Enabled

Ask: *Are we choosing the right tools for the job — and for our team?*

Not every AI tool is made for associations.
Don't chase shiny platforms. Choose tools that *amplify your strengths* and integrate with your current workflows.

You don't need the most complex system — you need the most useful one.

In your plan:
Have one tech goal per quarter. Keep it simple, achievable, and supportable. Don't launch three tools at once — that's a fast track to fatigue.

From Framework to Filter

The SMART AI Strategy™ isn't just something to write on the whiteboard at planning day.
It's a **lens** you apply to everything — every idea, every budget request, every process review.

So before anything makes it onto your 12-month plan, run it through this filter:

"Is this scalable, measurable, agile, relevant, and tech-enabled — or is it just exciting?"

Let that question save you from dozens of distractions and help you focus on *the few things that will make the biggest difference.*

The 12-Month Growth Cycle

Now that we've defined how to filter your decisions using the SMART AI Strategy™, it's time to build out what a full **12-month growth cycle** looks like in practice.

This isn't about overwhelming your team with too many projects or demanding results overnight.

It's about establishing a **repeatable rhythm** — a clear structure for moving from insight to action, and from small wins to big shifts.

This structure is broken into four quarters:

1. **Quarter 1: Start Small**
2. **Quarter 2: Measure & Learn**
3. **Quarter 3: Grow & Align**
4. **Quarter 4: Sustain & Systemise**

Each quarter has a focus — not a formula.
Every association is different, but the structure stays flexible and human-centred.

Let's break it down.

Quarter 1: Start Small

Theme: *Focus and Foundation*

This is your launch pad — not your leap.

Instead of setting 15 goals, choose **one or two high-impact, low-resistance areas** to explore AI's potential.
The goal is to build confidence — not complexity.

Priorities:

- Identify one clear member-facing or internal pain point.
- Choose one AI-enabled solution that addresses it (e.g. email summariser, chatbot, AI-assisted reporting)
- Run a pilot for 6–8 weeks.
- Build a small cross-functional team to test and reflect.
- Hold short debriefs: What worked? What didn't? What did we learn?

Leadership Role:

- Keep the scope narrow.
- Protect the team's capacity.
- Share progress with your board early (even if it's rough)

This quarter isn't about perfection — it's about *permission.* Permission to try. To learn. To lead.

Quarter 2: Measure & Learn

Theme: *Clarity and Reflection*

Now that you've got some initial runs on the board, it's time to build a feedback engine.
That means getting intentional about **measurement**, team reflection, and knowledge-sharing.

Priorities:

- Identify simple success metrics for your pilot (efficiency, engagement, feedback quality)

- Collect feedback from both staff and members.
- Introduce light dashboards or trackers.
- Run a "learning sprint" — a focused internal session to unpack insights.
- Share early wins at your next board meeting or strategy day.

Leadership Role:

- Model a "learn, not blame" culture.
- Keep conversations centred on *value*, not vanity metrics.
- Start teasing the idea of what could scale.

Quarter 2 is where belief starts to build.
Not because you're smashing goals — but because you're showing progress *with purpose.*

Quarter 3: Grow & Align

Theme: *Momentum and Maturity*

This is the pivot point.
You've tested. You've measured. Now you begin to **expand.**

But not everything needs to scale — only the projects that align with your strategy, your culture, and your capacity.

Priorities:

- Choose 1–2 successful pilots to scale across teams or departments.
- Document your learnings into a "What Works" guide.
- Sunset or pause projects that didn't serve you.
- Invest in small, targeted training to support capability-building.
- Map tech improvements to your member experience journey.

Leadership Role:

- Be bold enough to say *no* to projects that didn't deliver.
- Guide your board into strategic alignment with where growth is happening.
- Celebrate the team — reward impact, not just effort.

By the end of this quarter, your team should start to see that **AI isn't a project — it's a capability.**

Quarter 4: Sustain & Systemise

Theme: *Stability and Scalability*

This quarter is about locking in your gains.
It's about making sure what you've built is sustainable, supportable, and spreadable.

Priorities:

- Integrate successful AI tools into formal workflows.
- Assign owners and embed responsibilities into role descriptions.
- Create documentation and micro-training for onboarding.
- Update governance, risk, and procurement policies as needed.
- Design your growth plan for the *next* 12 months.

Leadership Role:

- Shift the narrative from "tech trial" to "business as usual."
- Position your team as a learning organisation.
- Begin sharing your journey externally.

This is where trust in the system gets built — and where your association moves from experimentation… to excellence.

Planning in Practice — The "Real-Life" Schedule

Let's move from strategy to story.

To truly see how the 12-month SMART AI Growth Plan works, we're going to walk through a **realistic, composite example** of how one association embedded AI into their year — one quarter at a time.

We'll call this organisation **The Allied Health Professionals Association (AHPA)**.
They're mid-sized, national, values-led, and like most associations — navigating:

- limited internal capacity.
- pressure to innovate.
- a legacy tech stack
- and a desire to better engage their members.

They didn't have a digital transformation team.
They didn't have a six-figure innovation budget.
They had *three core staff*, an overworked CEO, and a supportive but cautious board.

What they did have…,
A clear growth rhythm.

Quarter 1: Start Small

- **Pain Point Identified:** Members were overwhelmed by long, technical policy updates in the monthly newsletter.
- **AI Solution Chosen:** Use a free AI summarisation tool (like Claude or ChatGPT) to create member-friendly summaries.
- **Action:** A single team member tested the tool for 6 weeks, A/B testing subject lines and summaries.

- **Result:** Open rates increased by 23%.
  Member feedback was overwhelmingly positive — "Finally, I actually *read* the updates!"

This wasn't a huge leap.
But it was a **clear, measurable win** — and it sparked curiosity across the team.

Quarter 2: Measure & Learn

- **Reflection Held:** A 90-minute internal "what worked" session was held. Everyone brought one insight, one challenge, and one idea.
- **Dashboards Introduced:** The association used a basic spreadsheet to track engagement and efficiency metrics.
- **Team Learning:** Staff realised they were duplicating work across platforms — triggering a conversation about process automation.
- **Board Update:** The CEO presented the pilot outcome to the board, not as a finished project — but as an early success story.

Confidence started to build.
The narrative shifted from "Should we be doing AI?" to "Where else can this help?"

Quarter 3: Grow & Align

- **Expansion Area:** The team used AI to assist with:
  - drafting position statements
  - streamlining board reporting
  - pulling insights from member survey results
- **Capability Building:** A short AI upskilling session was held for all staff (internal, peer-led, 60 mins only).
- **Governance Note:** The board requested a basic AI usage policy and added "AI update" as a regular agenda item.

- **Member Benefit:** The policy team produced three new summaries with links to full reports, now being cited externally.

Momentum was now tangible.
The team wasn't chasing trends — they were solving real problems, with real-time learning baked in.

Quarter 4: Sustain & Systemise

- **Systems Update:** The email summarisation process was documented and handed to a new team member during onboarding.
- **Tech Audit:** Low-value tasks were flagged for potential automation in the next budget cycle.
- **Board Review:** AI efforts were reviewed alongside other KPIs — normalised as part of quarterly reporting.
- **Next Plan Created:** The CEO proposed the next 12-month roadmap, with three clear focus areas:
    - AI-enhanced member onboarding
    - Improved internal data workflows.
    - Scalable, AI-assisted CPD content curation

AHPA didn't become "an AI leader."
They became something far more valuable: **an agile, learning organisation.**

And in doing so, they built not just capability — but confidence.
Not just results — but rhythm.

Planning Day Prompt: Use the SMART Filter

When brainstorming initiatives for the year, run each idea through this checklist:

- **Scalable:** Can it grow without becoming chaotic?
- **Measurable:** Will we know if it's working?

- **Agile:** Can we pivot if we need to?
- **Relevant:** Does it help our members today?
- **Tech-Enabled:** Is it the right tool for the job?

Score each idea 1–5 on each criterion. Prioritise what scores highest overall.
Want a printable version for your next strategy day? Scan the QR at the end of the chapter.

The Mindset of Long-Term Transformation

Let's be clear: you don't need to get this perfect.

You don't need a bulletproof strategy.
You don't need to be a futurist.
You don't need to roll out five tech platforms in one year.

What you do need is a mindset — a way of thinking and leading that makes transformation not just possible, but **repeatable**.

This Isn't a Sprint — It's a Series

The greatest AI-enabled associations aren't the flashiest or the fastest.
They're the ones who **commit to the rhythm** — who start small, reflect often, and build systems that grow stronger over time.

Progress in this space isn't about having all the answers.
It's about having a pattern of asking better questions.

The 12-month SMART AI Growth Plan is not just a tool — it's a *lens*.
It helps you focus on what matters most and ignore the noise that pulls you off-track.

So if you're ever tempted to scrap your plan, or feel behind, or get discouraged by the pace — pause and remember:

- You're not late.
- You're not lacking.
- You're learning — and that's the work.

Transformation Lives in the Team

AI doesn't transform organisations.
People do.

All the dashboards in the world won't help if your team is burned out, afraid of change, or left out of the loop.

So as you implement this plan, keep the human front and centre:

- Create space for reflection and feedback.
- Celebrate small wins (especially the invisible ones)
- Let people experiment safely — and fail gracefully.
- Share ownership of what you're building together

Your people don't need perfection.
They need clarity. Encouragement. And a pathway forward they can see themselves in.

Annie's CEO Reflection:
"When I first led our team through a 12-month AI strategy cycle, I was terrified we were too behind to catch up. But what changed everything was consistency — not speed. We committed to one experiment, and the culture shifted. That's where real transformation begins — not with a tool, but with a team willing to try."

Don't Just Do AI — Be AI-Ready

This isn't about ticking boxes.
It's about becoming the kind of organisation that's *ready* — for whatever's next.

You can't predict every disruption.
But you can build the muscle to adapt with purpose.

That's what this plan offers:
A way to keep moving forward, no matter what's changing around you.

Keep Your Purpose Loud

Here's your final filter:
Does this help us serve our people better, in a way that reflects who we are?

If the answer is yes — go for it.
If not — park it.

Because the goal isn't to be the most "tech-forward" association in your sector.
The goal is to be the most **trusted, effective, and impactful** — using every smart tool available to make that happen.

Reflection Prompt:

Before you turn the page, take a moment with your leadership team to answer these:

1. What quarter of the 12-month growth cycle are we currently in?
2. Where do we feel strong? Where do we feel unsure?
3. What's one action we can take in the next 30 days to keep the rhythm going?
4. How will we protect our people while we build new processes?

Let's Keep the Momentum Moving

You've built something powerful.
Not just a strategy — a shift.

Now it's time to lock it in, lift it up, and use it to guide how you grow from here.

In the next chapter, we'll explore how to communicate this transformation — to your board, your members, your partners — so they don't just see the change…
They believe in it.

Let's make the invisible visible.

Author reflection

Looking back, I realise the biggest transformation wasn't in what we *did*—it was in how we *showed up* as a leadership team. We stopped looking for silver bullets and started designing rhythms. We stopped asking, "What tool should we use?" and started asking, "What system do we need?"
And slowly, those conversations changed us.
We stopped reacting to pressure—and started leading with purpose.
That's the magic of a real plan. It doesn't just guide your work. It reshapes your identity.

"Your legacy isn't defined by what you build, but by the impact you leave on those you serve." – Annie Gibbins

CHAPTER TWELVE

TELLING YOUR STORY

Sharing the "Why," Not Just the "What"

You've done the hard work.
You've shifted systems. You've empowered your team.
You've tested, tweaked, and implemented real, practical AI-powered changes.

But if you don't **communicate** that change — with vision, simplicity, and trust — people will either:

- Misunderstand it.
- Resist it.
- Or worse, *ignore it altogether.*

Because the truth is: people don't buy into a **tool**.
They buy into a **story** — one that shows them:

- Why this matters
- How it improves their lives
- And what it means for the future

This chapter is about **telling that story well**.

One of the biggest lessons I've learned as a CEO is that you can do all the right things operationally — but if people don't *understand* the why, change won't stick. I remember leading a major systems change that saved us hours each week — but members didn't see the benefit until we reframed the story around *their* outcomes. That's when everything shifted.

The Four Audiences You Must Bring Along

AI transformation in your association isn't a solo act. It's a collective shift — and like any shift, it succeeds or fails based on **who you bring with you, and how you bring them**.

Each group has a different lens, concern, and influence. If you want your AI momentum to last, you need to communicate **the right message to the right audience** — in their language, not yours.

Here are the four core groups:

1. Your Board: The Guardians of Trust and Risk

Boards don't fear technology — they fear **uncalculated risk** and unclear outcomes.
They want confidence that you're not chasing trends but building stability and future-readiness.

What to communicate:

- The strategic *why*: "This helps us serve our members better and stay relevant."
- The risk mitigation: "We've tested it. We're not scaling blindly."
- The measurable value: "We're saving time, improving engagement, and learning faster."
- The governance lens: "We're building AI use into policies, reporting, and decision-making."

How to engage:

- Present a short, clear AI update quarterly.
- Share what's working *and* what's still being explored.

- Link AI outcomes to strategic pillars (e.g. engagement, advocacy, education)

Pro tip:
Give your board members language they can use externally. Help them feel like champions of innovation — not just passive overseers.

2. Your Team: The Keepers of Culture and Execution

Your staff and volunteers are the ones living this change daily. If they don't understand it or feel overwhelmed, the transformation will quietly unravel beneath the surface.

What to communicate:

- Reassurance: "AI is here to support you — not replace you."
- Clarity: "Here's how these fits into your role and workflows."
- Inclusion: "Your feedback shaped this rollout. We're learning together."
- Empowerment: "You have permission to test, suggest, and grow."

How to engage:

- Offer bite-sized training or lunch & learn sessions.
- Celebrate team wins in applying AI.
- Invite input on where to improve.
- Make AI adoption a shared project — not a top-down mandate.

Pro tip:
Share behind-the-scenes stories. Let people hear each other's wins. It humanises the change — and builds buy-in from the inside out.

3. Your Members: The Reason You Exist

Members don't care if you're using AI.
They care if they're being heard, helped, and served.

That's why your messaging can't be "We've rolled out new AI tools."
It needs to sound like:

- "We've made it easier for you to find what you need."
- "We're responding faster and more personally."
- "You'll see better events, education, and support — thanks to smarter systems behind the scenes."

What to communicate:

- Benefits before tools
- Progress, not perfection
- Humanness first: "We're using tech to be *more* responsive — not robotic."

How to engage:

- Share mini "what's new" videos or updates.
- Highlight improvements in newsletters.
- Use stories and member testimonials to show the impact.
- Offer opportunities for feedback on changes.

Pro tip:
Ask: "What would make your experience feel smoother, faster, or more helpful?" Then show how you're using AI to deliver just that.

4. Your External Partners & Stakeholders

This group includes funders, regulators, allied associations, media, and sponsors.

They see your organisation as a thought leader — or at least, a player in a bigger ecosystem.

What to communicate:

- Your innovation narrative: "We're building a future-fit model for the sector."
- Your learning mindset: "We're testing, learning, and sharing."
- Your alignment to shared goals: "This helps us deliver stronger outcomes in [education, advocacy, policy impact]."

How to engage:

- Share your progress in reports, panels, or case studies.
- Speak about your approach at conferences.
- Offer to collaborate on pilot projects or cross-sector solutions.

Pro tip:
Position your association not just as a *user* of AI — but as a *voice of responsible, human-centred innovation* in your field.

A Final Word on Audience Alignment

You can't control how others perceive your transformation — but you *can* shape the story you're telling.

And when each audience sees themselves in that story — that's when trust builds, momentum spreads, and the change becomes permanent.

Crafting the Message — How to Tell the Right Story

Change isn't just about what you do. It's about **what people believe** you're doing — and why it matters.

When it comes to AI transformation, this means:

- *Avoiding the hype*
- *Staying human*
- *Making it relatable*
- *Speaking with clarity and care*

So how do you actually craft a message that lands — across internal and external audiences?

Start With the Why

Always begin by grounding the message in **purpose**.

LEADERSHIP REFLECTION

Ask yourself:

1. *What are we trying to improve for our members?*
2. *What problem are we solving?*
3. *How will this help people feel more supported, connected, and empowered?*

Example (internal staff update)
"We're using AI tools in our admin process so you can spend less time chasing documents and more time working with members."

Example (member communication)
"We've introduced a smarter way to personalise your experience, so you get updates that matter most to you."

Keep It Simple — Ditch the Jargon

Avoid phrases like:

- "Algorithmic workflow enhancements"
- "Generative pre-trained transformer integrations"

Instead, say:

- "We're using a smart tool to help us write reports faster."
- "You'll start getting emails that are shorter and more tailored to your needs."
- "We've tested a tool that can help summarise 20 pages into one clear paragraph."

Pro tip: If a 15-year-old couldn't explain it back to you, it's too complicated.

Tell a Story, Not Just a Status Update

Instead of:

"We're implementing AI tools in Q3 across membership and education."

Try:

"Last month, we tested a new AI-powered email summary tool with a small group of members. It took our staff 80% less time to produce, and more members clicked through. So we're expanding it this quarter — to save time and serve better."

Why this works:

- It shows momentum.
- It keeps it human.
- It builds confidence.

Case stories > Case studies.

Use the "3-Part Message" Framework

This is a quick, clear way to communicate your AI journey:

1. **Here's what we noticed:**
   "Our team was spending hours each week formatting member emails."
2. **Here's what we tried:**
   "We tested a simple AI tool to help us speed that up."
3. **Here's what happened:**
   "Emails now take 10 minutes instead of 2 hours — and member click rates have jumped."

This structure makes it easy for anyone — staff, board, or stakeholders — to understand both **logic and outcome**.

Give Your Team Talking Points

Don't assume people know what to say when asked, "What's this AI thing you're doing?"

Equip them with:

- A one-sentence summary
- A simple case example
- Reassurance that this isn't about cutting jobs — it's about improving service.

Example talking point:
"We're using AI in a few small areas — like member onboarding and event promotion — to make our work faster, more personal, and more impactful."

Own What You're Still Exploring

You don't need to have it all figured out. In fact, trying to pretend you do can create distrust.

Be transparent:

- "We're still learning."
- "We're starting small to make sure it works."
- "We'll share more as we grow."

This builds credibility — and it helps your community feel like **part of the journey**, not just recipients of it.

And if you're facing resistance — from within or from members — lean into it. Say, *"We know this is new. We're listening. Here's what we're learning."* Transparency breeds trust. And resistance often means people care — they just want to understand more.

Real Examples of What to Say — and What Not to Say

One of the fastest ways to build (or lose) trust during transformation is through your **words**.

Whether you're drafting a member update, briefing your board, or talking with media — your language signals what kind of leader you are and what kind of change this really is.

So let's compare.

INTERNAL TEAM EMAIL

Don't say:

"Our association is pursuing automation initiatives to streamline internal workflows using GPT-based API integrations."

Do say:

"We're trialling a smart writing tool to help reduce the time we spend drafting emails and reports — freeing us up for more member-facing work."

Why it works:
It's real, reassuring, and focused on people — not platforms.

BOARD UPDATE SLIDE

Don't say:

"AI technology rollouts are on track. No significant roadblocks."

Do say:

"In Q2, we successfully piloted two low-risk AI tools that cut admin time by 40%. We're reviewing feedback before expanding next quarter. No risks identified — and member response has been positive."

Why it works:
It's transparent, specific, and shows learning, not hype.

MEMBER NEWSLETTER

Don't say:

"We've implemented an AI framework to enhance communication delivery."

Do say:

"You might notice your emails from us are more personalised and to the point — we're using a new tool to help us tailor our updates to what you care about most."

Why it works:
It tells them what changed, how it helps them, and makes it feel human.

SPONSOR OR FUNDING REPORT

Don't say:

"Our transformation strategy includes scalable automation infrastructure."

Do say:

"We've identified three key areas where smart tools help us grow impact without growing budget — and we've built those into our member engagement plan for 2025."

Why it works:
It communicates ROI without drowning in tech lingo.

AT AN INDUSTRY PANEL OR CONFERENCE

Don't say:

"AI is transforming everything. It's complex, and we're still figuring it out."

Do say:

"We're focused on making AI work in practical, member-centred ways. We've started with small changes, learned what works, and now we're sharing what we've found — because this shift doesn't need to be scary."

Why it works:
It positions your association as relatable, competent, and generous — not vague or overwhelmed.

The Golden Rule: Focus on Benefits, Not Buzzwords

Whenever in doubt,

LEADERSHIP REFLECTION

Ask yourself:

"Does this sentence highlight a human benefit… or just a technology?"

If it's the latter — rewrite it.

Because real transformation isn't something you announce once.
It's a story you keep telling — one that evolves, includes others, and builds belief over time.

How to Train Your Leaders to Be the Voice of Change

Technology doesn't create culture.
People do.
And when you're leading AI transformation, it's not just about what *you* say — it's about what your **leaders echo, model, and embody** every single day.

Your board, senior team, regional reps, committee chairs — they are all *ambassadors* of your message.

If they hesitate, speak inconsistently, or contradict your vision, even unintentionally, it creates confusion.
But when they're aligned, clear, and confident?
Change spreads *with momentum* — not resistance.

Here's how to build that alignment and clarity, fast.

1. Start With a Shared Narrative

Create a one-page "AI transformation message guide" — and work through it together.

Include:

- **Why we're adopting AI** (in plain English)
- **What we've done so far** (with examples)
- **What members/stakeholders can expect**
- **What this means for our future**

Bring this to your next leadership session or board meeting. Workshop the language together. Let people ask questions. Give them space to share hesitations.

The goal: Co-create buy-in, not push a script.

2. Give Them Confidence, Not Just Talking Points

Most resistance to sharing a message comes from fear — fear of sounding out of depth, being asked tough questions, or misrepresenting the strategy.

Help your leaders feel confident by:

- Providing examples they can personalise.
- Practising short "elevator pitches" aloud
- Equipping them to answer common questions like:
    - *"Is this going to replace jobs?"*
    - *"How is member privacy being handled?"*
    - *"Why are we investing in this now?"*

Even a short internal training video or role-play session can work wonders.

3. Encourage Authenticity, Not Perfection

You don't need your leaders to sound like tech experts.
You need them to sound **like themselves — with clarity and conviction.**

It's okay if someone says:

"I was nervous about AI at first — but once I saw how it saved me 2 hours a week, I got it."

That's powerful. That's real. That builds trust.

Encourage personal reflection. Ask:

- What's *one thing* you've found helpful so far?
- What's *one way* you've explained this to someone else?

Those small moments are more persuasive than any polished deck.

4. Create Feedback Loops

Your leaders are your eyes and ears. Empower them to bring back:

- Questions they're being asked.
- Concerns they're hearing.
- Wins they're seeing.

This helps you refine messaging, strengthen training, and stay connected to what's happening on the ground.

Bonus: When people feel heard, they speak up more. That builds culture.

5. Celebrate Them Publicly

When a board member explains your AI shift at a conference...
When a team leader shares a win on LinkedIn...
When a chapter chair hosts an "AI and Associations" panel…

Celebrate it.

Even a simple "Shout-out to [Name] for championing our innovation journey!"
reinforces the message: **"This isn't my story — it's *ours*."**

Bottom line:
The strength of your strategy isn't just in your systems.
It's in the voices you amplify.

So train your leaders to be not just adopters of change — but **narrators** of it.

When to Speak Up — and When to Stay Quiet

In every AI journey, there are moments to amplify and moments to pause.

Knowing when to step forward with your message — and when to hold back until the story is clear — is what separates confident, credible leadership from rushed, reactive noise.

Let's break it down.

When to Speak Up

Speak early and confidently when:

You're testing something new — and want transparency.

"We're exploring a few new tools to help us better serve members. We'll share what we learn."

This shows proactivity, not secrecy. It builds trust.

You've had a win — no matter how small.

"We used a new tool to cut our event confirmation time from 3 days to 30 minutes — and members noticed."

Micro-successes are powerful momentum builders.

You're inviting participation.

"We'd love your input on how this could serve you better."

When members feel part of the journey, they're more likely to support the destination.

You're clarifying misinformation.

"There's been some confusion about what our new system does. Here's what it doesn't do — and what it's helping us achieve."

Get ahead of fear by telling the truth, simply.

When to Stay Quiet (For Now)

Sometimes the best communication is to **pause and prepare.**

Stay quiet (externally) when:

You're still in discovery mode — and don't yet know what works.

Don't announce a "transformation" when you're still in the sandbox.

Wait until you have one strong case study or example.

The message is too complex to explain in one sentence.

If it takes a 20-slide deck to say what's changing... your audience isn't ready, or the change isn't real yet.

Work on clarity *first* — before communicating broadly.

Internal alignment isn't there yet.

If your team or board is split, don't go public. Mixed messages create confusion and credibility damage.

Focus on internal education and unity first.

You're making a change that impacts jobs or long-standing processes — and haven't fully supported those affected.

Don't share a shiny "innovation story" externally if your staff feels blindsided or burnt out behind the scenes.

Get your house in order, then bring others in.

Golden Rule: Match Message to Maturity

Before you send anything out,

LEADERSHIP REFLECTION

Ask yourself: "Is the transformation far enough along to be felt, not just announced?"

If the answer is no — wait.
If the answer is yes — tell that story boldly.

Because timing isn't just a PR tactic. It's a leadership signal.

Tools, Templates, and Channels That Work

Now that we've covered what to say and when to say it…
Let's talk about **how** to say it — and what tools make it easy to stay clear, consistent, and human.

You don't need a communications degree or a full PR team. You just need a few well-used channels, repeatable tools, and simple rhythms.

Here's what works:

1. The AI Change Message Guide (Internal)

What it is:
A one-page cheat sheet your leaders and team can use.

What to include:

- "Here's why we're using AI."
- "Here's what we've done so far."
- "Here's what's coming next."
- A few key phrases everyone can use.

When to use:

- Team meetings
- Board briefings
- Town halls
- Staff onboarding

2. The "3-Part Story" Slide (Board + Stakeholders)

What it is:
A visual recap of a real AI use case, broken into:

1. The challenge
2. The tool we tested.
3. The outcome

Why it works:
Leaders want proof, not theory. Keep it short, real, and repeatable.

Template:

Challenge	Tool We Used	Outcome Achieved
Member response time was slow	AI reply suggestion tool	Replies now take 1 minute vs 10 — member satisfaction is up 12%

Use in:

- Strategic planning sessions
- Investor/funder reports
- Annual reviews

3. "What's New + Why It Matters" Email (Members)

Keep a standing format to share progress without overwhelm:

Subject line:

What's new at [Your Association Name] — and why it matters to you

Body:

- A 2–3 sentence update
- One bullet list of member benefits
- Optional: short testimonial or quote from a team member

Repeat monthly or quarterly. Members will start *looking forward* to seeing what you're building.

4. "Behind the Build" Updates (Optional Blog, Video or Podcast)

Pull back the curtain.
Show members and stakeholders that AI isn't just a tech thing — it's a human thing, driven by thoughtful leadership.

Ideas:

- Staff spotlight: "How I use AI to save time every day."
- Member Q&A: "Your top 5 questions about our AI rollout — answered."
- Micro-cast: 3-minute audio updates featuring your CEO or innovation lead.

These don't need to be polished. They just need to be **real.**

5. Amplify Wins Across Channels

Whenever something works, **share it in multiple places:**

- Add it to the website's "What's New" section.
- Include it in a visual on LinkedIn.
- Mention it at the next town hall.
- Let your board tell it at their next sector roundtable.

Rule of thumb:
If it mattered once, it's worth repeating *at least* 3 times in different places.

Wrapping Up

You don't need fancy campaigns.
You don't need marketing degrees.
You need **honest stories, clear tools, and human delivery**.

That's how trust is built.
That's how innovation gets adopted.

And that's how your AI journey becomes more than a system shift —
It becomes a *movement* your whole community believes in.

Quick Recap: Making Your AI Story Stick

Step	Action
1	Know your four audiences (Board, Team, Members, External)
2	Use clear, benefit-driven language (ditch the jargon!)
3	Share stories of progress, not perfection
4	Give leaders tools to speak confidently
5	Choose when to speak — and when to wait
6	Repeat success stories across 3+ channels

Keep this on your desk or share it with your comms team — it's the heartbeat of every AI update that works.

One of the most powerful lessons I've learned. When leaders speak from clarity and care — not just strategy — people follow. I've seen hesitant teams become bold adopters, and sceptical boards become innovation champions. Not because the tech was perfect, but because they were brought along for the ride. Your voice matters more than any tool you implement. Use it wisely.

In the next chapter, we'll explore how to **measure what matters** — because when it comes to AI, transformation is only as powerful as your ability to *track progress, reflect honestly, and evolve wisely.*

CHAPTER THIRTEEN

FUTURE-READY LEADERSHIP STARTS WITH YOU

If you've reached this chapter, something profound has clicked into place.

You see it clearly now:

AI isn’t a threat to your leadership—
it’s a tool to amplify it.

It’s not about becoming more technical—
it’s about becoming more intentional.

This isn’t simply a conclusion—it’s the bridge from insight to action. It’s your moment to pause, reflect deeply, and prepare to lead boldly. This chapter is your rallying cry, a manifesto, and my personal invitation to you to step forward with confidence, clarity, and courage.

The 10 Truths of a Future-Fit CEO

1. **You don't need all the answers.**
   You need better questions, smarter experiments, and a willingness to learn publicly.
2. **Your true power isn't control—it's clarity.**
   The sharper your vision and values, the easier it becomes to adapt and inspire.
3. **Culture isn't a side-note—it's your multiplier.**
   Accelerating broken systems multiplies dysfunction. Cultivate culture deliberately.
4. **Your board can be your greatest ally—or your greatest obstacle.**
   Guide them to possibility using a language they trust and understand.
5. **Leadership happens publicly and privately.**
   Your posts may inspire, but your daily actions shape true influence.
6. **You're more than a membership organisation.**
   You're a learning ecosystem, a trust engine, and a movement maker.
7. **The technology will evolve endlessly.**
   Your focus should remain firmly on the why—and who—you serve.
8. **Innovation isn't an event—it's a rhythm.**
   Make innovation a monthly practice, a culture of ongoing exploration and growth.
9. **AI won't replace you—but future-ready leaders using AI effectively will replace those who don't.**
   Lead from a position of clarity and courage, not caution.
10. **Your legacy doesn't begin at the end—it starts today.**
    It's crafted moment by moment, decision by decision, choice by choice.

A Personal Letter from Annie

Dear Leader,

You are exactly where you need to be.

I didn't write this book because I have all the answers. I wrote it because I've sat exactly where you're sitting right now—in those challenging board meetings, navigating fatigue, uncertainty, and quiet yet powerful ambition.

I've seen your struggle. I've felt your fears. But more importantly, I've recognised your potential—the extraordinary potential you might sometimes overlook when you're in the thick of it.

Consider this book your guide and companion: not a prescriptive rulebook, but a supportive voice saying, "Yes, you can do this." You have what it takes to transform your association without losing its soul.

Every tool, framework, example, and reflection in these pages has been carefully designed to remind you of something crucial:

You are ready.
You are capable.
Your leadership matters now more than ever.

So, embrace AI not as a distant trend, but as your greatest ally—a tool that amplifies your vision, your voice, and your humanity.

Let clarity lead you through complexity.
Let courage carry you through uncertainty.
Let your leadership become the beacon that others follow.

I'm so proud of how far you've come.

Now, go lead—and show your organisation what's truly possible.

Warmest regards,
Annie

The Smart Association Manifesto

Print this. Share this. Live this.

- We lead from purpose, never from panic.
- We prioritise clarity over control.
- We don't wait for permission—we adapt proactively.
- We build trust before technology.
- We simplify first, then systemise.
- We test bravely, learn humbly, and grow consistently.
- We speak human, never hype.
- We are bold, not perfect.
- We leverage AI, but always lead with heart.
- We are creating the future our members deserve.

Ready for Your Next Step?

Your journey doesn't end here. It accelerates.

To continue your momentum and turn insight into impact, download your FREE AI Leadership Toolkit today:

Appendix A: The SMART AI Strategy™

Your Roadmap to Intentional AI Integration

Introducing AI into your association isn't about chasing tech trends—it's about empowering your mission with clear, confident, and strategic decisions. The **SMART AI Strategy™** is your go-to framework for ensuring every AI initiative aligns precisely with your organisation's purpose, capacity, and values.

No overwhelm.
No guessing.
Just clarity, alignment, and impact.

What SMART Stands For

Use these five simple questions to filter every AI opportunity, idea, or tool:

Letter	Focus Area	Guiding Question
S	**Scalable**	"Can this solution scale smoothly without draining resources?"
M	**Measurable**	"Are there clear metrics we can track to assess success?"
A	**Agile**	"Can we rapidly test, learn, and adapt without disruption?"
R	**Relevant**	"Does this directly address what our members or staff truly value?"

Letter	Focus Area	Guiding Question
T	**Tech-Enabled (Not Tech-Obsessed)**	"Does this technology genuinely enhance our mission, rather than complicating it?"

When to Use the SMART Model

Integrate the SMART criteria into your strategic discussions, especially when:

- Considering new AI tools or platforms
- Pitching AI initiatives to your board or stakeholders
- Running agile pilot projects
- Evaluating ongoing AI performance
- Refining and recalibrating your overall digital strategy

Think of SMART as your strategic compass—guiding confident, data-informed decisions every time.

SMART AI Strategy™ in Action: An Example

Scenario:
Your association wants to use AI-powered onboarding automation to streamline member experiences.

Here's how you'd apply SMART:

Letter	How to Apply SMART	Practical Considerations
S	**Scalable**	Can our onboarding automation seamlessly support a growing membership?

Letter	How to Apply SMART	Practical Considerations
M	**Measurable**	How will we quantify improvements (e.g., hours saved, higher retention rates)?
A	**Agile**	Can we initially test with a smaller member group (pilot) before full rollout?
R	**Relevant**	Is onboarding automation directly solving a known pain point for new members and staff?
T	**Tech-Enabled**	Does AI-enhanced onboarding genuinely improve our member experience without sacrificing personal interaction?

Real-World Outcomes: Using SMART

Leaders who've adopted the SMART AI Strategy™ consistently report:

- **Higher Board Confidence:** Clearer criteria mean faster approvals and stronger support.
- **Reduced Team Overwhelm:** Focusing on agile, scalable solutions prevents resource burnout.
- **Better Member Experiences:** Technology aligned to member values creates deeper engagement.
- **Stronger Organisational Clarity:** Fewer distractions, more intentional growth.

Take the Next Step

Assess your next AI idea or project with confidence using the **SMART AI Strategy Scorecard**. This practical tool lets you score ideas quickly and clearly, ensuring they align with your strategic priorities.

Appendix B: Pilot Project Planner

Start Small. Learn Fast. Scale Smart.

AI initiatives don't need huge budgets or high risks. The best innovations often begin as lean, agile pilots designed to test ideas quickly, build team confidence, and deliver tangible results.

This planner provides a structured yet simple approach to designing, executing, and scaling your first (or next) AI pilot project—without overwhelm.

STEP 1: Define Your Purpose

Clearly articulate the specific problem you're trying to solve. Don't start with technology—start with the impact you want.

Examples:

- "We spend over 12 hours per week manually creating member communications."
- "Event registrations have plateaued; we want to test AI-driven personalisation."

Guiding Question:
"What measurable improvement do we aim to achieve?"

STEP 2: Apply the SMART AI Strategy™ Filter.

Evaluate your idea against these key criteria:

Criteria	Question to ask	Your Notes
Scalable	Can we easily grow this if it succeeds?	
Measurable	Are clear success metrics available?	
Agile	Can we test this idea quickly (within 2–6 weeks)?	
Relevant	Does it directly support our members or strategic goals?	
Tech-Enabled	Will technology meaningfully improve this?	

Tip:
If you answer "No" to two or more criteria, rethink your scope or choose a different project.

STEP 3: Assemble Your Lean Pilot Team

Keep your team small, agile, and accountable:

Role	Name	Responsibility
Pilot Champion		Overall leadership, communication, and accountability
Tech Lead (if needed)		Ensures tech tools integrate smoothly and trains the team

Role	Name	Responsibility
Key Stakeholder(s)		Provides insight, feedback, and validation

Tip:
Limit your core pilot team to no more than 3–4 people to keep things agile and efficient.

STEP 4: Set Clear Success Metrics

Define exactly how you'll measure success. Choose at least one quantitative and one qualitative metric.

Metric Category	Your Metric Example	Target
Efficiency	"Hours saved per week"	From 12h to 3h/week
Member Satisfaction	"Positive member feedback"	80%+ rate experience as 4+/5
Engagement	"Email open rates increase"	Improve open rates by 25%
Quality/Accuracy	"Error rate reduction"	Reduce processing errors by 50%

Tip:
Clearly defined, measurable metrics make board reporting and future decision-making easier.

STEP 5: Pilot Execution & Feedback Loop

Run your pilot for 2–6 weeks. During the pilot period, intentionally:

- **Monitor Results:** Track metrics weekly, adjust tactics if necessary.
- **Capture Feedback:** Regularly solicit input from your team and members.
- **Document Learnings:** Clearly record what's working and any unexpected challenges.

Weekly Log Template (simple):

Week	Metric Achieved	Insights Learned	Challenges Faced
Week 1			
Week 2			
Week 3			

Tip:
Keep logs simple and narrative-based, suitable for sharing internally and externally.

STEP 6: Review, Reflect, and Decide

Post-pilot, bring your team together and answer these questions honestly:

- What did we learn from this pilot?
- What adjustments would strengthen this approach?

- Is this initiative worth expanding or scaling?
- Should we refine this further or try something else?

Then make a clear decision:

| ☐ **Expand/Scale** | ☐ **Refine Further** | ☐ **Pause/Pivot** |

Tip:
If your pilot succeeded, share results widely—celebrating your team's achievement builds enthusiasm and buy-in.

Pilot Project Summary Template (Example):

Project Title	**"AI-Driven Member Communication"**
Duration	4 weeks
Core Team	CEO, Tech Lead, Membership Officer
Challenge	12h/week wasted on manual emails
AI Solution	ChatGPT-assisted content automation
Results	Cut email drafting time from 12h/week to 3h/week; Member satisfaction rose by 22%
Next Steps	Scale pilot to other communication areas

Appendix C: AI Readiness Scorecard

How Future-Fit Is Your Association, Really?

Your association's future isn't defined by technology alone—it's shaped by your mindset, systems, and strategic approach to change.

Before you dive into large-scale AI initiatives, this practical scorecard gives you a clear snapshot of your current strengths and critical areas needing attention.

Use it to prioritise your next moves, align your team, and ensure your association is genuinely ready to lead confidently in the age of AI.

What You'll Gain:

- A realistic snapshot of your association's AI readiness
- Clear identification of strengths and weaknesses
- Alignment within your board and leadership team
- A roadmap to prioritise immediate, high-impact actions.

Use this as a self-assessment or complete it collaboratively with your board, executive team, or an innovation working group.

The 5 Dimensions of AI Readiness

Rate your association on a scale from **1 (Not True at All)** to **5 (Consistently True)**.

1. Leadership Alignment

Statement	Score (1–5)
Our executive team actively seeks and discusses emerging technology opportunities.	
AI and digital innovation are regular topics in our strategic conversations.	
Our board understands the strategic role of technology in achieving our mission and serving members.	
Leadership actively supports and funds agile experimentation and pilot projects.	
Subtotal:	___ / 20

2. Operational Systems

Statement	Score (1–5)
Our digital systems integrate smoothly, allowing seamless data-sharing across departments.	
We've clearly identified repetitive tasks ripe for automation or AI improvement.	
Staff can reliably access the most up-to-date documents, tools, and information.	
We can test new digital tools without disrupting day-to-day operations.	
Subtotal:	___ / 20

3. Member Experience

Statement	Score (1–5)
We regularly identify areas where our member experience could improve.	
Member communications feel personal, timely, and consistently relevant.	
Our digital platforms are intuitive, accessible, and consistently user-friendly.	
We regularly gather actionable feedback and consistently use it to improve.	
Subtotal:	___ / 20

4. Data & Insight

Statement	Score (1–5)
Our data is accurate, up-to-date, and easily accessible to relevant teams.	
We regularly track and analyse member engagement trends and behaviours.	
Strategic decisions routinely draw from data and insights, not assumptions or gut feeling.	
Staff can easily interpret and use data without needing specialised data analysts for basic tasks.	
Subtotal:	___ / 20

5. Culture & Capability

Statement	Score (1–5)
Our team demonstrates curiosity, openness, and enthusiasm for trying new technologies and methods.	
We actively encourage and provide resources for experimentation and innovation.	
Staff feel psychologically safe proposing new ideas, tools, and improvements.	
Learning, innovation, and outcomes are visibly celebrated, not just busy work.	
Subtotal:	___ / 20

Your AI Readiness Score: _____ / 100

Score Range	What It Means	Recommended Next Steps
80–100	**Future-Fit Leader**: Your association is highly ready for advanced AI integration.	Prioritise scaling successful pilots and share leadership insights across your sector.
60–79	**AI-Aware & Growing**: Good foundation, but some gaps still exist.	Focus on refining alignment, strengthening systems, and delivering early wins to build momentum.
40–59	**At-Risk & Reactive**: Potential is clear, but immediate attention is required.	Prioritise critical systems upgrades, targeted cultural initiatives, and start small-scale pilot projects immediately.

Score Range	What It Means	Recommended Next Steps
Below 40	**Urgently Needs Attention**: Your current approach risks your association's relevance.	Immediately address leadership alignment, mindset shifts, and initiate carefully chosen, low-risk pilot projects to build confidence.

Where to From Here?

Now that you've assessed your AI readiness, use this scorecard as a strategic conversation starter with your board or leadership team. Highlight priorities, agree on actions, and build internal momentum together.

Remember:
Clarity drives action. Action drives change.

Your association's future starts right here—today.

Appendix D: Glossary of Terms

Cut Through the Jargon. Speak AI with Confidence.

AI doesn't need to feel intimidating. This glossary simplifies common AI and digital transformation terms into plain language, empowering CEOs, board members, and your entire team to engage confidently and clearly in conversations around AI.

Think of this as your go-to reference guide—clear enough for newcomers, detailed enough for strategic discussions.

AI & Machine Learning Terms

- **Artificial Intelligence (AI)**
  Technology that enables machines or software to mimic human capabilities such as problem-solving, learning, planning, and decision-making.
- **Machine Learning (ML)**
  A subset of AI focused on algorithms that improve automatically through experience (data), becoming more accurate without explicit programming for each scenario.
- **Deep Learning**
  An advanced type of ML, using layered neural networks to process vast amounts of data, leading to breakthroughs in voice recognition, image analysis, and complex decision-making.
- **Neural Network**
  A computer system modeled loosely after the human brain's interconnected neurons, used in deep learning to recognise patterns, interpret data, and make predictions.
- **Natural Language Processing (NLP)**
  Technology enabling computers to understand, interpret, and generate human language. Commonly used in chatbots, virtual assistants, and text analysis.

- **Generative AI**
  AI capable of creating original content—text, images, music, or videos—based on learned patterns from existing datasets. (e.g., ChatGPT, DALL·E)
- **Chatbot**
  An AI-powered conversational agent designed to answer questions, provide information, and assist users in real-time, enhancing member engagement.
- **Large Language Model (LLM)**
  A sophisticated AI model trained on extensive textual data to generate human-like conversational responses. (e.g., OpenAI's GPT-4)
- **Algorithm**
  A sequence of computational steps or rules followed by software or AI systems to solve problems or perform tasks efficiently.

Operational & Implementation Terms

- **Automation**
  Leveraging technology or AI to complete repetitive, manual tasks automatically, freeing up human resources for strategic or creative work.
- **Workflow**
  Defined sequences of tasks and actions your association uses to execute projects or processes efficiently. Identifying workflows is critical for successful AI integration.
- **Pilot Project**
  A controlled, small-scale implementation to test and evaluate the potential benefits, challenges, and ROI of a new AI solution before scaling it organisation-wide.
- **Proof of Concept (POC)**
  A simplified demonstration or experiment designed to prove the feasibility and practical potential of an AI idea or technology before investing fully.
- **Prompt Engineering**
  The practice of carefully crafting questions, instructions,

or inputs given to AI systems to achieve optimal, accurate, and relevant outputs.

- **Tech Stack**
  All technology solutions, platforms, and software applications your organisation uses, including CRM systems, membership databases, marketing tools, and communication platforms.

Data, Analytics & Insights Terms

- **Data-Driven Decision Making**
  An approach to making informed strategic choices based on objective data analysis rather than intuition or assumption.
- **Insight**
  A meaningful, actionable discovery gained from analysing data, driving better-informed strategies and decisions.
- **Dashboards**
  Interactive visual displays that summarise real-time data and analytics through graphs, charts, or visualisations, allowing leaders quick insight into organisational performance.
- **Segmentation**
  Dividing your members into clearly defined groups or segments based on demographics, interests, behaviours, or engagement, enabling targeted, personalised interactions.
- **Predictive Analytics**
  Using historical data, statistics, and AI to anticipate future outcomes, trends, and member behaviours, allowing proactive strategies and responses.
- **Data Governance**
  Policies, standards, and processes ensuring your data's accuracy, security, and compliance, essential for ethical and responsible AI deployment.

Strategic Leadership Terms

- **AI Readiness**
  Your association's preparedness—in culture, skills, technology, and strategy—to successfully implement, leverage, and benefit from AI tools.
- **SMART AI Strategy™**
  Annie Gibbins' practical and proven strategic framework designed to ensure AI adoption is **Scalable, Measurable, Agile, Relevant, and Tech-enabled**.
- **Digital Transformation**
  The process of using digital technologies—including AI—to fundamentally improve organisational operations, culture, member experiences, and strategic positioning.
- **Innovation Culture**
  A work environment that actively encourages experimentation, creativity, learning from failure, and openness to new ideas—including AI and digital technologies.
- **Change Management**
  A structured approach to supporting individuals, teams, and entire organisations in adapting smoothly and effectively to new technologies, processes, and strategies.
- **Ethical AI**
  AI designed and implemented according to clearly defined ethical standards—ensuring fairness, accountability, transparency, inclusivity, and respect for privacy.

Security & Risk Terms

- **Cybersecurity**
  Measures, practices, and technologies designed to protect your association's digital assets, data, and operations from cyber threats and attacks.

- **Data Privacy**
  Practices, regulations, and technologies that protect your members' and organisation's sensitive personal information, ensuring compliance with laws like GDPR and Australia's Privacy Act.
- **AI Ethics Charter**
  A formal document outlining your organisation's guiding principles, policies, and standards for ethically responsible AI use, providing transparency and accountability.
- **Risk Management Framework**
  A systematic approach used by associations to identify, evaluate, and mitigate potential risks associated with adopting and using AI technologies.

Additional Business & Membership Terms

- **CRM (Customer Relationship Management)**
  Software systems used to manage your association's interactions and relationships with current and potential members, optimised further with AI-driven personalisation.
- **Member Experience (MX)**
  The sum of all interactions and experiences a member has with your association, greatly enhanced through targeted AI-driven insights and engagement strategies.
- **SaaS (Software as a Service)**
  Cloud-based software accessed via subscription rather than owned outright—commonly employed by associations for flexibility, cost-effectiveness, and ease of scalability.

Appendix E: Further Reading & Resources

Keep Learning. Keep Leading. Keep Evolving.

AI leadership isn't a destination—it's an ongoing journey. This carefully curated collection includes the most authoritative, practical, and insightful resources for association leaders committed to continuous growth and future-focused leadership.

Consider this your personal resource library, available whenever you need fresh insights, practical tools, or inspiration.

Books for Strategic Thinkers

AI & Innovation

- **The Age of AI and Our Human Future** *(Kissinger, Schmidt & Huttenlocher)*
  Thoughtful perspectives on how AI will shape human society, geopolitics, and leadership.
- **Human + Machine: Reimagining Work in the Age of AI** *(Daugherty & Wilson)*
  Explore how AI can augment, rather than automate away, the human workforce.
- **Artificial Intelligence and Machine Learning for Business** *(Steven Finlay)*
  Clear, jargon-free insights for executives looking to grasp practical AI applications.

Leadership & Organisational Culture

- **Start with Why** *(Simon Sinek)*
  Essential for purpose-driven leaders looking to anchor AI strategies in meaningful human purpose.
- **Dare to Lead** *(Brené Brown)*
  A transformative guide to leading courageously and vulnerably through technological change.

- **Atomic Habits** *(James Clear)*
  Valuable for embedding small, impactful changes in organisational behaviour and culture.

Future-Ready Associations

- **Race for Relevance** *(Coerver & Byers)*
  Timeless strategies on maintaining relevance in rapidly evolving membership organisations.
- **The End of Membership as We Know It** *(Sarah Sladek)*
  A provocative look at adapting association structures and strategies for the next generation.

Insightful Podcasts

- **HBR IdeaCast (Harvard Business Review)**
  Groundbreaking interviews on leadership, AI, innovation, and strategy.
- **The AI Podcast (NVIDIA)**
  Conversations with global leaders pioneering practical AI solutions.
- **Leading Learning Podcast**
  Valuable insights for association executives, covering tech trends, member engagement, and digital transformation.
- **Nonprofits Are Messy (Joan Garry)**
  Honest discussions about innovation, culture, and tech adoption in associations and nonprofits.
- **The Naked CEO Podcast (Annie Gibbins)**
  Practical leadership strategies and candid conversations for association executives navigating digital transformation.

High-Impact Websites & Blogs

- **Harvard Business Review (HBR.org)**
  Industry-leading articles on leadership, AI, and business transformation.
- **World Economic Forum (weforum.org)**
  Global AI trends, ethical implications, and innovative leadership frameworks.
- **McKinsey Insights (mckinsey.com)**
  Deep research and thought leadership on AI implementation and leadership strategies.
- **NTEN.org**
  Dedicated resources on ethical and practical AI integration specifically for associations.
- **SmartAssociations.ai** *(Founded by Annie Gibbins)*
  Tailored insights and strategies designed specifically for associations navigating AI adoption.

Professional Development & Courses

- **MIT Sloan Executive Education**
  Intensive short courses on AI, digital transformation, and leadership.
- **Stanford Online – AI Business Programs**
  Practical and prestigious courses for executives aiming to deepen their AI expertise.
- **Women in AI (womeninai.co)**
  Supporting inclusive leadership and gender diversity in AI through events and training.
- **Lead With Annie (leadwithannie.com)**
  Personalised mentoring and targeted workshops designed exclusively for association leaders.

Tools & Platforms to Explore

- **ChatGPT / OpenAI**
  Generative AI tools for brainstorming, research, content creation, and ideation.

- **Canva AI Tools**
  Intuitive visual design and AI-supported content generation for busy teams.
- **Notion AI**
  Enhanced productivity through AI-driven note-taking, project management, and task automation.
- **Otter.ai & Tactiq.io**
  Instant meeting transcriptions and actionable summaries, dramatically reducing admin overhead.
- **FutureTools.io**
  Extensive directory of vetted AI applications for every organisational need.

Industry Reports & White Papers

- **Deloitte Annual Tech Trends**
  A comprehensive annual guide outlining emerging AI and technology trends affecting leadership.
- **State of AI Report (McKinsey & Company)**
  In-depth analysis and forecasts that provide strategic clarity for forward-looking association leaders.
- **NTEN's AI for Nonprofits Report**
  Insights specific to AI adoption challenges and opportunities within the nonprofit sector.

Leadership in this age demands continuous evolution—not just of technology, but of our mindsets, skills, and approaches. I encourage you to bookmark, explore, and revisit these resources frequently. Stay curious, stay courageous, and keep moving forward.

Let's keep the conversation going! Connect with me personally to continue your leadership journey.

With gratitude,
Annie Gibbins

ABOUT THE AUTHOR

Annie Gibbins
CEO | Global AI Mentor | Speaker | Author | Change Catalyst

Annie Gibbins is not just a CEO — she's a catalyst for transformation.

With over 25 years at the helm of purpose-led organisations across health, education, and traditional medicine, Annie Gibbins has become a trusted global voice on leading with clarity, courage, and strategic innovation in times of rapid change. From traditional sectors to emerging spaces like AI-enabled leadership, Annie brings bold vision and practical wisdom to every boardroom she enters.

As the CEO of the **Australian Traditional Medicine Society** and Director of the **Australasian College of Optical Dispensing**, Annie lives the realities of association leadership: tight budgets, legacy systems, staff fatigue, and the pressure to stay relevant. She's walked the walk — and now she mentors leaders worldwide to walk their own path with greater confidence, clarity, and purpose.

Annie is the founder of **SmartAssociations.ai** and **Lead With Annie**, guiding association executives, CEOs, and boards through the complexities of digital transformation. She is also a

Grand Stevie® Award Winner for Women in Business and the powerhouse behind **Women's Biz Global** — a multi-award-winning platform supporting women in building authority, publishing boldly, and leading fearlessly.

As a **#1 bestselling author** and publisher, Annie has written and released multiple influential titles, including:

- *TWINS: A Memoir of Hope and Resilience*
- *A Woman's Guide to Business Domination*
- *Heart Warrior*
- *Memoirs of Successful Women*
- *Awaken Your Divine Feminine*

Annie's signature style is personal, powerful, and refreshingly straight-talking. Whether she's on stage, leading a strategic retreat, hosting *Women's Biz Leaders TV Show* and *The Naked CEO* podcast, she cuts through the noise to deliver insights that stick — and strategies that scale.

She believes that **AI doesn't replace the human element — it amplifies it.** And her mission is to help leaders like you use AI not as a buzzword, but as a bridge to deeper connection, smarter systems, and legacy-driven leadership.

When she's not mentoring, writing, or leading transformation, Annie can be found at sunrise with a strong coffee, a big idea, and a heart for service.

To reach out to Annie personally, email:

annie@anniegibbins.com
leadwithannie.com | smartassociations.ai

Made in the USA
Monee, IL
05 November 2026

67383966R00184